GW01184645

Microcar *Mania*

Chris Rees

Microcar *Mania*

THE DEFINITIVE HISTORY OF THE SMALL CAR

bmp

PUBLISHERS BOOKMARQUE TYPESETTERS
PUBLISHING
Minster Lovell & New Yatt · Oxfordshire

First published 1995
© Chris Rees, 1995

(The author asserts his moral rights)

All rights reserved. No part of this book may be reproduced or transmitted in any form or by any means, electronic or mechanical, including photocopying, recording or by an information storage or retrieval system, without prior written permission of the publisher or copyright holder.

ISBN 1-870519-18-3
(limited casebound edition)
ISBN 1-870519-38-8
(collectors' edition)

....

British Library Cataloguing in Publication Data

A catalogue record for this book is available from the British Library

Set in 10 on 12 point Times
Typesetting and origination by Bookmarque
Printed on Fineblade Smooth 115 gsm
Published by Bookmarque Publishing · Minster Lovell & New Yatt · Oxon
Printed and bound by Butler & Tanner Ltd · Frome · Somerset

CONTENTS

PHOTOGRAPHS & CAPTIONS IN THIS BOOK

Where 2 or more photographs appear on a page, the captions (in descending order) relate to the photographs on a left to right basis.

ACKNOWLEDGEMENTS

THANK you to all the people who have helped out in making this book possible: in particular, to Gary Axon, Scott Brownley of BMW (GB), Stephen Boyd of the Scootacar Register, Stan Cornock of the Bond Owners' Club, Phil Boothroyd of the Messer-schmitt Owners Club, Mike Shepherd, Dave Simonetti, Darren Styles, Nigel Russell and especially Jean and Edwin Hammond of the Register of Unusual Microcars.

PREFACE

THE publication of this book fills me with great pleasure as it has allowed me to fill one large gap in the motoring bibliography. The world of microcars is both diverse and fascinating from all points of view, but little recognition has been afforded these little cars. Very few books have ever been published about the smallest cars ever made. Microcars have been uniformly neglected by the classic car press on the grounds that they do not sell magazines. However, this does at last appear to be changing because of the fast growing interest in the subject.

For many of the same reasons, the publication of a definitive history of microcars is also fraught with difficulty. Quite apart from the huge number and considerable obscurity of microcar marques, the size of a book which attempted to be comprehensive would be massive; that means a very high cover price out of the range of the most enthusiasts' pockets. There are several high-quality, detailed books on microcars published in German but even these (which generally concentrate only on micros of the 1950s) carry an elevated price tag.

So this book is intended as an affordable, and collectable, international review of microcars from 1940 to the present day, detailing the most important and most interesting of the many hundreds of microcars made over the last 50 years. Regrettably, it cannot be encylopaedic.

However, for the sake of completeness, included at the end of the book is an A-Z of Microcars which lists every microcar marque and every model of which I personally have knowledge. One thing is certain: it is not, and can never be, 100% complete. I do welcome any correspondence regarding any additions to, or errors found in, the A-Z section. Who knows, perhaps this may form the basis towards a true Encyclopaedia of Microcars in the future!

Chris Rees

INTRODUCTION

"HOW very great are the very small," said Thackeray. His dictum might just as well apply to the motorised contraptions we call cars as to the luminaries of wisdom. For the very smallest of all forms of motor car are, in many ways, indeed marvellous.

The average enthusiast, let alone the average motorist, has been apt to poo-pooh the microcar throughout its history, or at least be mystified about its *raison d'etre*. The common conception is of an unspeakably awful car to drive, bottom-wobblingly crude and uncompromisingly ugly. In short, its attractions, as well as its dimensions, are microscopic.

Such philosophies are born of ignorance. They ignore the microcar's reason for coming into existence in the first place. Microcars responded to the pressing needs of their day: offering 'motoring for the masses' in the case of the cyclecars of 1910-25; overcoming raw materials shortages in the case of the post-war bubble cars; and alleviating the pressures of chokingly congested traffic and dwindling fuel reserves of modern times.

Nearly always, the design of a microcar calls for some ingenuity. True, there have been truly bad examples of design in the microcar breed, as there are examples of poor design in all other fields. Within the constraints imposed by the closely confined parameters of the microcar, the engineers behind them have often produced brilliant results.

What exactly is a microcar? Everyone has a different idea. The term 'microcar', which has become the accepted description of the smallest class of cars ever made, obviously has no lower limit. Anything with more wheels than a moped and a seat to plonk a bottom on counts as a car. There are several possible 'classes' of microcar, from three-wheeled mopeds through 'scooter-cars' (the 'rollermobil' or 'cabin scooter') up to what amounts to large cars in miniature. What are the upper limits? It's difficult to draw precise boundaries but, as a guide, anything above 10ft in length is bigger than a Mini — which is definitely not a microcar, but a 'mini car'.

But you will find some cars over 13 feet in length mentioned. Above all, it is engine size which determines a microcar. Most of the cars in this book have engines below 500 cc, but the unofficial 'limit' set by the Register of Unusual Microcars in Britain is 700 cc. That's the figure which I have taken as the prime determinant for inclusion in *Microcar Mania*, although there are a few exceptions.

There is some evidence that attitudes towards microcars are changing. They have become very much more popular in recent years, even collectable. Yesterday's unwanted reminder of rations and shortages are today's rare classics. Certainly in classic car circles, the microcar is now regarded as a prized object — rather to the surprise of most microcar enthusiasts. After the recent classic car price crash, the only cars to retain and positively *increase* their value have been microcars — to the extent that a pristine Messerschmitt Tiger is worth more than a pristine Rolls-Royce of the same period!

Certainly there is more appreciation of what microcars offer. Naturally you have frugality, ease of repair and restoration, superb abilities in traffic, no parking problems and even a certain chic. But the most important attribute of any true microcar is what it *does not* have. Endearment comes about in the realisation that these cars get by without mechanical complexity or preconceptions as to the relative 'cool' of a particular model. In a world of add-ons and hype, they excel by being the minimum. In my years with Citroën 2CVs (although not perhaps true microcars), I often thought what a good idea it would be for everyone to be forced to drive one for a week: I learned a hell of a lot about driving by not having big power reserves to draw on, like the revelation that it is quite possible to keep up with all but the most maniacal of GTi drivers in most modern road conditions simply by technique alone.

The support for microcar owners and enthusiasts today is unprecedented. In Britain there are several one-make clubs catering for the Messerschmitt, Heinkel, Isetta, Scootacar, Bond, Reliant and Nobel. And there is the celebrated Register of Unusual Microcars which encompasses the whole spectrum. The club scene is also big in Europe, with particularly strong clubs in Germany, Holland and Sweden.

There are museums devoted to microcars (most notably the Story Museum in northern Germany) and one is in preparation by the Register of Unusual Microcars in Britain. Plenty of other museums recognise the significance of the microcar in the wider perspective of motoring history with numerous exhibits.

All writers studying the microcar phenomenon end their analyses with a phrase like: "Surely the microcar represents the future of motoring". Personally I don't subscribe to that view, for there will always be the need to transport families and luggage around, requiring vehicles much more ample than the most capacious microcar. And there will always be, I think, a 'thundering chariot' mentality which is basically incompatible with the microcar.

But the inevitability of small cars becoming more important players in the motoring world is inescapable. One has only to look at what most of the Japanese manufacturers are turning out these days to realise that sophistication and integrity need not be strange bed-partners with the microcar. Nor indeed enjoyment, style and comfort.

To illustrate this book some pictures were of dubious quality but have been included to make the book as complete as possible.

CHAPTER ONE

CYCLECARS & THE ORIGINS OF THE MICROCAR

THE very first motor cars ever made could be described as microcars. The car which started it all off, the 1885 Benz, was a simple three-wheeler with a one-cylinder engine developing 0.75 bhp. The recipe for these pioneers was to produce the simplest workable machine possible.

There were numerous firms in the early days interested in the production of an internal combustion vehicle and their first cars were almost always compact, straightforward, small-engined vehicles — the very definition of a microcar. Leon Bollee created his first small car in 1895 and, as he was the first to do so, he had to give it a name and coined the phrase 'voiturette'. His three-wheeler had a 650 cc single-cylinder engine which developed 3 bhp.

By the time most makers actually came to producing a vehicle for sale, there was really only one type of customer: the wealthy. They called for something more than just a means of transportation and most cars of the Edwardian age were mostly motorised carriages. Even Leon Bollee had abandoned the voiturette by 1903 and was building a large, expensive car.

But there were still advocators of the small car — and indeed have been ever since. Humber, Lagonda and Sunbeam all began their careers with small cars in Britain, while in France, Renault and De Dion were at the forefront of voiturette developments.

In America, there were numerous attempts to market cars known as 'buckboards', among the starkest and most uncomfortable means of transport ever devised. These consisted of a wooden board with wheels attached, seats plonked on top and an engine usually lashed directly to the back axle. They achieved some popularity and were even made as late as the 1950s.

The Edwardian voiturettes and tri-cars were very marginal in general terms. It was not until about 1910 that small cars received a boost with the arrival of the cyclecar. These placed more emphasis on light weight, combining motorcycle and car technologies, and attempted to provide for

'New Motorists' — probably the first attempt at making the motor car truly popular (the Ford Model T was just beginning to be manufactured on a production line at this time).

As opposed to the impractical tri-cars of the early 1900s, the cycle-car followed a definite pattern: usually quite long, narrow cars, often with tandem seating, almost invariably four-wheeled, using small engines and simple mechanicals. They were cheap — sometimes as little as £50 — and could often return over 50 mpg. More importantly, they were much more easily fixable than any other type of car.

And indeed the cyclecar did become popular for a while. Dozens of marques flourished in Europe and the USA, bringing cheap motoring to the masses.

But the advent of the First World War dealt the breed a blow from which it never really recovered. The many manufacturers fluttered quickly out. Some reappeared after the war, but the golden age of the cyclecar — from about 1910 to 1914 — had passed.

There was another reason for the cyclecar's decline: mass produced cars were getting cheaper to make and the Ford Model T was technically aeons ahead of the mostly primitive cyclecar offerings, and was cheap to buy. But the Austin Seven of 1922 was the car which truly sealed the fate of the cyclecar.

Austin's Seven brought large-car features to a small car and was justifiably extremely popular. As production increased, its price progressively lowered until it bowed under £100, by which time it was impossible to offer very much for significantly less money. Some remnants of the cycle-car boom did echo on through the 1920s.

Only the best marques survived intact offering something unique in the 1930s: these were the three-wheelers of Morgan, BSA, Raleigh *et al* and broadly split into two camps: the ultra-basic, ultra-cheap family car and the sporting three-wheeler.

Abroad, where there was not such fierce competition from a small 'real' car like the Austin, there were several firms which lasted through the 1930s, such as the Neander and Standard Superior in Germany and the Mochet in France.

The 'buckboard' car was a peculiarly American phenomenon, made all the more peculiar by the fact that it was utterly unsuited to a life of driving on rough American roads. Most consisted of a few planks of wood, some seats, bicycle wheels and a tiny engine fitted directly to the rear axle or, in one case, attached to a trailing fifth wheel. This example, an Auto Red Bug of the late 1920s, ran on electric power.

But by the time hostilities broke out in 1939, there were pitifully few such manufacturers left and all of them disappeared during the war. Some, like the French Mochet and English Lloyd reappeared after the war, but the whole scene was rather different in those difficult years following the Second World War. For the birth of the 'new cyclecars', or microcars, heralded an unprecedented boom — indeed the like of which has never been seen since. The microcar had truly arrived: for which, see Chapter 2.

In Britain, the AC name is well-known. Its origins stretch as far back as 1908, when John Weller built his first Autocarrier (the initials of which spelt AC). The production model, named the AC Sociable, was a single-cylinder three-wheeler with basic tiller steering costing under £100. Mostly bought by tradesmen, it remained on sale until 1914, long enough to have established AC as a major manufacturer of cars.

The GN was the archetypal British cyclecar and one of the most popular of the breed. Combining simplicity with very low prices, the GN might justly be described as the first practical 'car for the masses', production having started at Hendon, Middlesex, in 1911. The cars were extremely light (only 6.5 cwt) and were consequently attractive to competition-minded customers.
By 1925, however, like most of its kind, the GN had died.

A vicar's son, H.F.S. Morgan produced an early and definitive cyclecar which became the most popular and most respected of its genre. From 1910, the Morgan three-wheeler was built in Malvern (where it remains even today, though not making cyclecars!). Family versions and some spectacular sporting models were made and, unique in the history of cyclecars, three-wheeled Morgans were built right up until 1952.

The car which killed the cyclecar in Britain: the Austin Seven. With a 747 cc engine producing 13 bhp, it offered levels of performance and sophistication never seen in cyclecars at a very cheap price. It was the king of baby cars, offering everything you might expect from a larger car, simply on a smaller scale. The Seven lasted 17 years until 1939.

Even Bugatti attempted a microcar. Ettore had designed the Bebe Peugeot in 1912, but eschewed cyclecars. His Type 56 was a cheeky little electric car used to ferry customers around his factory and did not enter production. Strangely, Bugatti's first car following the outbreak of World War 2 (in 1942) was a tiny doorless prototype called the Type 68 (see p. 150). Its four-cylinder 369 cc engine developed 15 bhp. There was also a 1:1 scale model of a two-door coupé version of the 68 shown as late as the Paris Salon of 1948, but neither model reached production. Ettore Bugatti died in 1947.

Fiat's 500 Topolino was a masterstroke of design. It arrived in 1936 and offered synchromesh for the gearbox, independent front suspension and hydraulic brakes in a pretty two-seater rolltop body. The 570 cc side-valve engine could power the 500 to a top speed of 55 mph at 55 mpg and cost only £120 in Britain. It developed in only minor bursts until its demise in 1954. Pictured is a 1949 500C.

CHAPTER TWO

THE BUBBLE BOOM

TO most, the true 'golden age' of the microcar was in the 20 years following the end of the Second World War. The question of providing popular transport in the post-war years was heartily addressed by a large band of engineers across Europe — and even in America — by the creation of innumerable small cars.

The problem was that motoring in any form was very difficult following the war. If you were lucky enough to own a car, petrol was in short supply and was rationed in the UK well into the 1950s, so it was beneficial if you had the most economical car possible.

Most civilians had no car, however. Second-hand pre-war bangers fetched ridiculous prices. Metals were in such restricted supply that new car production was very limited; those that were built usually went to priority classes, like doctors. In France, Citroën even had a policy of supplying its 1948 2CV only to deserving customers — and the waiting list was years long. There was a yawning gap populated by millions of potential drivers who would snap up anything that was available to them.

It was in this environment of deprivation that the microcar flourished. It satisfied the need for cheap, practical and economical transport. More than this, the microcar was usually an easy object to manufacture and used few materials. Indeed, with the arrival of glassfibre as a workable substance for bodywork from about 1948, there was really little need to use much metal in manufacturing.

In the years immediately following the war, supply shortages stopped even the most basic forms of transport from being produced. Only a handful of the models made during the war — notably some of the many electric microcars in France — continued in production and there were very few new cars (although there were some notable exceptions, like the Fend in Germany and Rovin in France).

It wasn't until the start of the 1950s, as the severest shortages began to ease, that potential manufacturers overcame the difficulties. Very quickly, all over Europe, a new breed of tiny cars mushroomed. Especially in Germany, where supply shortages were chronic, the profusion was mind-boggling. Microcars achieved popularity in France, Spain, Italy and Britain. In part, British demand blossomed at the expense of the motorcycle-and-sidecar combination, whose popularity as family transport waned dramatically during the 1950s. British drivers also benefited from the tax and licensing concessions granted to three-wheelers weighing less than 8 cwt.

The rich variety of cars which ensued included some brilliant designs like the Messerschmitt and the Berkeley. It also brought about some of the most absurd creations ever seen on the road. One thing was definite: there was little conformity. Each country developed its own styles and each manufacturer adopted different approaches. Some tried to make large cars in miniature; others adapted motorcycle technology; still others created bold and original machines specifically tailored to meet the demands of microcar drivers.

One of these was the Iso Isetta of 1953. This was the original 'bubble car', which spawned a whole gamut of similar looking spheroids. The commercial success of the bubble car, particularly in Germany and Britain — and of course its arresting appearace — has led to it becoming synonymous with the microcar as a breed. But the vast majority of microcars of the time were, however, distinctly un-bubble-like in appearance.

The momentum of the microcar movement was given a huge push with the first oil crisis of 1956, when the Suez canal became engulfed by war. Once again, fuel was in short supply and demand for small cars shot up: the bubble car truly experienced a boom.

But just as the cyclecar had been rendered obsolete by the Austin Seven in 1922, so the bubble car was burst by the new Austin Seven of 1959. Alec Issigonis's Mini was specifically born of Leonard Lord's dislike for bubble cars. There were even rumours that BMC put pressure on firms which supplied components to microcar firms to stop doing so, just prior to the launch of the Mini. Had the Mini not been so brilliant, perhaps the microcar might have lasted longer. But it was brilliant, succeeding in offering a quantum leap of attractions over other small cars of the period. Compared to most microcars — and indeed many larger cars — it was capacious, fast and fun, yet only slightly less frugal and not significantly more expensive. In any case, prosperity was assuredly returning to the western world and there were no longer problems about fuel or material supplies: people realised they could afford to own a larger car.

Slowly the micro lights of Europe began to go out. In Germany, the boom was effectively over by the early 1960s; in Britain, it echoed on with decreasing intensity until the mid-1960s; in certain countries, like Greece, the last of the 1950s microcar boom straggled on as late as the 1970s.

But the microcar did not die: there were notable survivors, such as Bond and Reliant, which developed the theme in the ensuing years and a new breed of microcars began to be born: for this part of the story, see Chapter 3.

GREAT BRITAIN
LARMAR

The Cyclopean eye of the Larmar was but a prelude to the horrors which trailed it.

The Essex-based firm, Larmar Engineering, produced a single-seater primarily for invalids from 1946. It was a frankly hideous contraption whose absurd proportions were enforced by the designer's wish that the car should be able to fit through a doorway: hence it had a width of just 28.5 ins (72.5 cm). It had such refinements as a cyclops headlamp, folding hood and a windscreen. Its 249 cc BSA single-cylinder engine sat in the rear and drove only one of the rear wheels by chain to a top speed of 35 mph (56 kh/h). A slightly improved version, offered with a 350 cc engine, was also made up until 1951, when the Larmar's slender appeal had been well and truly overtaken by the times.

An early Bond Minicar looks dwarfed in front of the much expanded Bond Mk F.

BOND

Alongside Egon Brütsch, Lawrie Bond was one of the most prolific post-war microcar designers. Not only did he found his own marque — one of the first British post-war ventures — he designed or had a hand in several other microcars, most notably the Berkeley and the Opperman.

In 1948, Lawrie Bond had developed an extremely basic prototype as a form of cheap, popular transport. Wanting very much to put it into production, he began talks with Sharp's Commercials of Preston, Lancs and eventually they agreed to make the car under Bond's name.

So the Bond Minicar was launched in 1949. One can realise just how basic the prototype was from the 'improved' specification of the aluminium-bodied production car: it had cable and bobbin steering, a Perspex windscreen, open bodywork, no doors, brakes on the rear only and no rear suspension (the tyres were left to absorb all the bumps). A 122 cc single-cylinder two-stroke Villiers engine sat immediately in front of the single front wheel, driving it by chain through a three-speed gearbox.

These early Bonds were intended merely as runabouts but, such was the success of the model, many people wanted to use their Minicars as all-purpose transport. So Bond also offered a larger 197 cc Villiers engined version.

An improved model, the Minicar Mark B, arrived in 1951 with the luxury of rear suspension, but no dampers! Like the Mark A, most production cars were two-seaters, although vans, Minitrucks and 2+2 Family Safety Saloons were also offered.

A door appeared as standard for the first time on the Mark C of 1952 — on the passenger's side only. There was also now a front brake and better suspension. In appearance, the Bond also got front wings, making it look like a much larger car. One might dismiss this as pure puff, but the arrangement allowed the Bond for the first time to do a now-celebrated manoeuvre: the front wheel and engine could turn through 180 degrees, allowing the Bond to turn within its own length. By 1955, Bond was making nearly 300 cars per month.

The Mk C Bond Minicar could even be pressed into a service as a four-seater ...just!

With its engine mounted on the front wheel and a cavernous space surrounding it, servicing a Bond simply meant climbing inside...

The Mark D (1956) shared the C's body but added 12 volt electrics. But the Mark E of 1957 marked a complete design overhaul. Now fitted with a chassis (as opposed to the stressed body with strengthening cross-members of the earlier models), the Bond was now strong enough to be fitted with full-size doors — on both sides of the car! The body looked much different, sporting a slab-sided profile and squarer lines.

As the weight of these more sophisticated Bonds was rising, Villiers was approached to build a bigger engine. Reboring the existing unit to 246 cc provided an extra 4 bhp and a 25% increase in performance; cars fitted with this engine were known as the Mark F. A four-speed 'box was now standard.

In 1961, the Mark G added a few more refinements: a hard-top with a 'breezeway' rear window, larger (10 inch) wheels, hydraulic brakes and wind-up windows. An estate version (which Bond claimed to be "the world's first three-wheeled estate car") arrived shortly after, of which there was a Ranger van version.

The Minicar had been a very successful design — probably the most popular of any British microcar. But sales of the still very basic and slow car were declining so that, by 1966, production of the Minicar ceased. A total of just under 25,000 of all types had been built.

Since 1963, Bond had also been making a Triumph Herald based sports car called the Equipe. The experience gained in making its glassfibre body was put to use in the Minicar's 'replacement', the 875 of 1965 — a very different car.

For the 875 and Bug stories, see Chapter 3.

The curious Allard Clipper, introduced at a time when Allard's sports cars were facing tough times.

ALLARD

Alongside AC, Reliant and Bond, Allard was one of the few firms engaged in that peculiarly English practice of making microcars and sports GT cars both at the same time. In Allard's case, most opinion remains that they should have stuck to building sports cars.

The Clipper was very weird by any standards but it was also desperately crude. Introduced in October 1953, it was an oddly shaped glassfibre hardtop coupé — one of the first ever cars fitted with a plastic body. It was designed by David Gottlieb (who was later to design the Powerdrive) and looked like a pair of eggs given a styling job by Chevrolet. On some versions, there was even a dickey seat in the rear 'boot'.

The Clipper was a light car at just 6 cwt (305 kg). The engine was a 346 cc Villiers 8 bhp unit mounted in the rear and driving only one of the rear wheels. There were cooling problems and a driveshaft weakness, making it woefully unreliable.

The project lasted for only two years or so before Sidney Allard pulled the plug on the project to concentrate on his sports cars, which were also then having trouble selling. Estimates of the number of Clippers made vary from four to 12 to 40, but today only three are known to survive.

RODLEY

If its cardboard-box appearance didn't put the car buyer of 1954 off, the Rodley's disposition towards self-incinerating surely would have. The decision by Rodley Automobiles of Leeds to mount a 750 cc JAP engine in the completely enclosed rear end led to drastic cooling problems and fires were not unheard of. It was optimistically described as a four-seater coupé, would apparently do 75 mph and was to be made at the rate of 50 a week. But no more than a dozen were built by the time the Rodley ducked out in 1956.

AC

AC Cars is one of the longest-established of all British car firms, yet has consistently remained in a penumbra. Its first forays from 1908 concerned the lightweight Auto Carrier Sociable (see Chapter 1). AC made more of a name for itself with the large sports cars and grand tourers it produced after the Second World War, including the legendary Cobra.

But AC never lost its microcar roots. In the 1950s, it met the needs of the day with the AC Petite, which was an earnest attempt to make a comfortable economy car with three wheels.

Hardly a bastion of aesthetics, the Rodley had a reputation for self-combusting due to its cooling problems.

The AC Petite's appeal was almost entirely towards the economy side of the motoring equation, although it was more comfortable and spacious than most.

It arrived in 1953, a squarish two-to-three-seater with a steel and aluminium body which, from a distance, looked like it might have four wheels. Closer inspection revealed that under the front lay a single wheel which steered the car. The engine, a single-cylinder 346 cc Villiers industrial unit, sat in the back. A 50 mph (80 km/h) top speed was claimed.

The Mk 2 version arrived in 1955, incorporating a slightly larger (353 cc) Villiers engine. Whereas the Mk 1 had a smaller front wheel than those at the rear, the Mk 2 had the same size wheels all round.

Compared to other microcars, the AC Petite was comfortable and civilised. But it was hardly pretty, charismatic or, for that matter, much of a driving experience. Ending its production life in 1958, around 4,000 cars were built.

GORDON

If followers of the football pools were looking to hit the jackpot with the microcar introduced on the profits Vernon Industries made out of them, they were sorely mistaken. The Gordon was in no respect a happy car.

Vernons had built an invalid carriage called the Vi-Car since 1952 and used this as the basis of its Gordon three-wheeler, which entered production in 1954.

In form, the Gordon looked like many other British microcars of the period: ugly. But it was also crude *in extremis*. The standard-issue Villiers 197 cc one-cylinder engine actually sat outside the main bodywork in a little 'pouch' of its own on the offside, in front of the rear wheel. Quite apart from what this must have done for its handling, it meant that there was only one door to get in by — on the passenger's side. The steel bodywork (all 11 feet 2 inches of it) featured a folding roof.

Neither was the Gordon a great performer: a top speed of 45 mph (72 km/h) was quoted and the car was too heavy for much acceleration. Drive was transmitted to only one of the rear wheels by chain. The only selling-points for the Gordon were that it was simple, cheap to buy and cheap to run: indeed it was Britain's cheapest car, costing just over £300. But it was hard to believe Vernons' claim that it was "Britain's finest three-wheeler family car". Between 1954 and 1958, it succeeded in winning over only a few hundred customers. There are only four known survivors.

It is hard to believe that the Gordon was Britain's finest anything, *but it was at least extremely cheap.*

Another primitive microcar of the 1950s: the Atom from Fairthorpe. It used a BSA motorbike engine sited in the rear.

FAIRTHORPE

Fairthorpe is a company unique in the history of British specialist cars. It began life in 1954 from a base in Chalfont St Peter with microcars and went on to create unusual and, it must be admitted, crude sports cars right up until 1978.

Fairthorpe was the product of Air Vice-Marshal Donald 'Pathfinder' Bennett whose first design was the Atom. On paper, it had quite an advanced specification for a microcar of its era: a tubular steel backbone chassis with independent suspension all round, hydraulic brakes and a choice of rear-mounted engines: 250 cc BSA single-cylinder (Mk 1), 322 cc Anzani twin (Mk 2), 350 cc BSA (Mk 2A) and even a corking 646 cc BSA twin (Mk 3). In this latter form, the Atom could reach 75 mph (as opposed to 45 mph for the 250 cc version).

However, the reality was not so advanced. The 2+2 GRP body may have looked aerodynamic, but it was also pig ugly (especially in prototype form with free-standing headlamps) and had claustrophobically small doors and windows. It was also unbearably crude to drive.

Improvements were made: extra windows helped visibility, the headlamps became faired-in and there was even a convertible option. Fairthorpe sold 44 examples by the time it was replaced by the Atomota in 1957/58.

The Atomota was a rather different car, having a front-mounted engine (still the BSA 650), live hypoid rear axle, coil springs at the rear end and a synchromesh four-speed gearbox. It looked similar to the Atom but sported a whacking pair of fins at the back — a concession to the styles of the age. Later literature referred to the model as the Atom Major.

It was sold complete or in kit form but with little success: only a handful were made by 1960, when Fairthorpe abandoned microcars in favour of its larger-engined sports cars, the Electron (1956), Electron Minor (1957) and Zeta (1960).

OPPERMAN

Having witnessed Lawrie Bond's Minicar become Britain's best-selling microcar, the tractor manufacturer Opperman sought out his services to design them a new microcar. Bond duly created the Unicar in 1956, a car which was to combine "big car comfort with small car economy."

The Unicar was a conventional looking machine with a steel-reinforced glassfibre body seating 2+2. Its 328 cc Excelsior engine was placed in between the two rear seats, just in front of the rear axle (a mid-engined microcar!), where it was unfortunately prone to overheating. Its 18 bhp could take the car to 60 mph. The two rear wheels were placed close together, avoiding the need for a differential.

To buy new, it cost £399 10s or, from 1958, you could build one yourself and save 33% on purchase tax. From 1956 to 1959, some 200 Unicars were sold.

Opperman's new model for 1958 looked all set to be a big success. The

The Unicar made by Opperman was a rare example of a microcar also sold in kit form.

Stirling 2+2 coupé looked smart with its professionally-styled GRP bodywork. It shared the same mechanical layout as the Unicar but had a wider rear track and a differential. Its engine was the 424 cc Excelsior Talisman 25 bhp twin, but a second (and final) prototype was built in conjunction with Steyr-Puch of Austria with a 493 cc flat-twin engine from the Austrian version of the Fiat 500.

However, that is as far as the project went. There were some rumours that BMC were concerned that the Stirling might be too close to its forthcoming Mini for comfort and put pressure on component suppliers not to deal with Opperman, but whatever the cause that was the end of the Opperman micros. A handful of Unicars and a single Stirling survive today.

Opperman's pretty Stirling coupé would surely have been a success had it entered production.

The MVM was one of several British attempts to make a micro-sports car but was Guernsey's sole motor manufacturer.

MVM

Guernsey's one and only car manufacturer should have been Leslie Le Tissier's Manor View Motors (MVM). In 1956, it made a small open sports two-to-three-seater with a ladder-frame chassis, all-independent suspension and a glassfibre body. The two-cylinder two-stroke 325 cc Anzani 18 bhp engine was centrally mounted and a top speed of 60 mph (96 km/h) was claimed. It was intended that the MVM would be sold exclusively in mainland Britain but in the event only two prototypes were ever built.

BERKELEY

Lawrie Bond, the prodigious microcar designer, approached Charles Panter of Berkeley Coachworks in 1955 with the idea that he should build his idea for a micro sports car. Previously, Berkeley had pioneered the art of glassfibre as early as 1948 in its caravans. Panter agreed and Bond built three prototypes in 1956. The result was the tiny (10ft 2in) car of GRP unitary construction and independent suspension all round, weighing just 5.5 cwt. Even with the tiny 322 cc Anzani engine (15 bhp), it was a nippy little car. Known as the Sports SA322, it was an instant success. But supply problems with the Anzani led Panter to source another engine, the 328 cc Excelsior unit, and the model was renamed the SE328 (later B65). An excellent performer (top speed 65 mph), with keen handling and braking, it won significant praise. But Bond's chassis could easily handle more power, so the B90 was launched in 1958 with a 492 cc three-cylinder version of the Excelsior unit and a four-speed 'box instead of three speeds. Berkeley's numbering system, incidentally, was based on its models' top speeds: it followed that the B90 could achieve 90 mph. Both two and four-seat versions were offered, the latter known as the Foursome.

In the search for an engine more suited to its American market, Panter fitted his first four-stroke to a Berkeley, the Royal Enfield 692 cc twin in 1959. With two states of tune churning out 40 and 50 bhp, the two new models were known as the B95 and B105 respectively. They could be distinguished by their heavier nose treatment incorporating a rectangular grille.

Panter decided to modify the rear end of the B65 and create a three-wheeler called the T60, launched in October 1959. Why it took him so

This early Berkeley was one of the first cars to be made with a glassfibre reinforced Bakelite plastic body.

long is a mystery, but it was an instant smash hit, knocking most other micro three-wheelers for six. Using the 328 cc Excelsior engine, it notched up an incredible 1,850 sales in just over a year; the four-wheelers had sold just over 2,310 in total.

By the end of 1960, it was all over for Berkeley. The caravan market had collapsed and dragged Berkeley's car side with it. Berkeley's potential saviour, the larger and more conventional Bandit, arrived too late to save the sinking enterprise.

In 1991, an enthusiast by the name of Argyle began offering replica Berkeley T60 kits based on Mini mechanicals.

In an extended wheelbase, four passengers could just squeeze into the Berkeley Foursome.

Berkeley's T60 three-wheeler is sometimes said to be a failure: in fact, it was easily the firm's best-selling model.

FRISKY

The purple racing colours of Egypt were only ever painted on one car: the Phoenix sports car of 1956. It was the brainchild of British businessman Captain Raymond Flowers who was then based in Cairo. At the same time as the sports racer, he also developed a microcar of very strange appearance (featuring a single up-and-over door) but that was never marketed. Flowers was forced to return to Britain in the middle of the Suez crisis.

Back in Britain, he decided to push ahead with developing his Phoenix

The first Frisky prototype had unusual gullwing doors and was optimistically described as a five-seater.

microcar. He collaborated with Henry Meadows Ltd of Wolverhampton (then making engines), who eventually built the production cars. When it reappeared in Britain in 1957, it sported very much more attractive lines penned by Michelotti and now had gullwing doors. The new car was claimed to offer seating for five and was given the name Frisky.

However, this was judged an impractical production proposition and the Frisky was restyled with open bodywork, conventional doors and sharper lines (which even ended in tail fins!). The glassfibre body was made in the nearby Guy lorry works and was mounted atop a separate ladder chassis. The rear wheels were set so close together that no differential was required.

The 249 cc Villiers two-stroke engine drove the rear wheels by chain through a motorbike gearbox; reverse was obtained by reversing the engine. The front suspension was Dubonnet-type independent. Very shortly after launch, the 324 cc Villiers 16 bhp engine was also offered as an option.

Called the Friskysport, the little car really lived up to its name: it could manage 65 mph (104 km/h) while returning 56 mpg. Both coupé and convertible versions were sold.

This did well enough to finance the development of another project, the Friskysprint, in 1958. This was a dramatic-looking open car designed by Gordon Bedson and was much more of a sports car. It measured only 37.5 ins (95 cm) high. Its 492 cc three-cylinder two-stroke Excelsior engine booted out 30 bhp and Meadows claimed it would do 85 mph (135 km/h). But it never came to production and the project was sold to an Australian firm which later marketed the car as the Zeta Sports (see Chapter 3).

Meanwhile, Captain Flowers decided to press ahead with a more basic Frisky, which he called the Family Three. This was a three-wheeler with accommodation for four people, using 197 cc, then 250 cc two-cylinder engines. Production concentrated on three-wheelers after 1961. A larger version, the Prince, was offered from 1960 with a 324 cc or 328 cc engine. The only other Frisky was also called the Frisky Sport. Not much is known about this model, which was a larger and more conventional-looking car. Whether it went into production is not known. After its fourth successive change of premises, Meadows abandoned production of the Frisky in 1964.

A rather different appearance for the restyled production Friskysport, available with three or four wheels.

The so-called Frisky Sport appeared in the firm's early catalogues but production did not begin.

PEEL

Despite its small size, the Isle of Man-based firm Peel Engineering was one of the fountains of the microcar garden, forever producing fresh ideas and new designs. Its main business was making glassfibre moulds for motorcycle fairings, boat hulls and so on, in which it was something of a pioneer.

In 1955, it was encouraged by the appearance of several GRP-bodied cars to enter the world of motor car manufacture. Its first project was the Manxman, a small three-wheeler with a steel tube chassis and an enclosed GRP body. Initially a rear-mounted 350 cc Anzani two-stroke twin drove the single rear wheel by chain, but later a 250 cc Anzani unit was fitted. A top speed of 50 mph (80 km/h) was claimed with a fuel consumption of 90 mpg.

Two hammock seats were supplemented by a rear compartment which could carry two children or 16 cubic feet of luggage. Entry was gained via a most unusual system: the semi-circular door pivoted at its rear base, swinging upwards and back, flush with the bodywork.

The plan was to sell the 4.5 cwt Manxman in kit form for £299 10s, but it is very unlikely that any were made. However, Peel did make several examples of its larger GRP bodyshell for Ford Ten chassis, called the P1000, which appeared in the late 1950s.

The car for which Peel is now most famous is the P50 of 1962. This was almost without doubt the world's smallest ever passenger car. In Peel's literature, it stated that the prototype measured just 4ft 2ins (127 cm) long, 2ft 9ins (84 cm) wide, and 3ft 10ins (117 cm) tall. Production cars were 4ft 5ins (135 cm) long and 3ft 3ins (99 cm) wide, and weighed a mere 132 lbs (60 kg).

In prototype form, it appeared with a single front wheel and twin rear wheels, but production models had the layout reversed. The P50 consisted of a single seat surrounded by a one-piece GRP shell incorporating a single front headlamp. The DKW 49 cc fan-cooled engine sat underneath the driver, where it was extremely noisy. It drove the single rear by chain via a 3-speed gearbox.

The P50 was hardly a comfortable machine to drive and there were stabilising 'nodules' to the rear of the bodywork on some cars to stop them toppling over on corners. The provision for reversing consisted of a handle fitted on the back, the owner being expected to manhandle the car into a parking space. No two Peels were ever identical, being individually finished at the factory.

Still, at a price of £199 10s in 1963, many were tempted. New P50s were despatched in a wooden box which could double up as a garage! Up to 1966, somewhere in the region of 75 cars were delivered, many of which still survive today.

Peel's next model was the Trident three-wheeler of 1965, which was larger than the P50, but only just (measuring 6 feet or 183 cm long). The GRP body of the Trident consisted almost entirely of a forward-hinging

The Isle of Man-built Peel P50 easily scuttles off with the accolade of 'the world's smallest car'.

section which had a transparent Perspex dome as a roof and a flat glass windscreen. The umbrella handle-shaped steering column was also hinged and rose with the canopy.

The Trident could seat two side by side (just) or some were supplied as single-seaters with a shopping basket beside. The engine was initially the same 49 cc unit as the P50, but some cars had 100 cc Vespa engines. It was kick-started using a detachable pedal. With a turning circle of 8 feet, there was really no need for a reverse gear, which is just as well since none was fitted.

Costing £189 19s 6d, the Trident was even cheaper and marginally more practical than the P50. Peel advertised it as being "almost cheaper than walking", having a consumption of about 100 mpg. A four-wheeled electric version also appeared in 1966, offering an unusually good range and brisk acceleration.

Peel stopped making the Trident in 1966, after selling about 45. By this time it had also sold on its Mini-based GRP coupé body project, called the Viking Minisport, to Bill Last (who ironically also made a GT car called the Trident). Peel had created all sorts of other car projects in glassfibre, including a jeep-type vehicle measuring only 87 inches long and an all-GRP Mini replica shell, but these never became true production models.

TOURETTE

Knowledgeable visitors to the 1956 Earls Court Motor Show might have thought that they'd come across the Brütsch Zwerg when they arrive at the Progress Supreme stand. The egg-shaped Tourette on display there looked remarkably similar — but it was, in fact, an original design.

Having built some glassfibre-bodied scooters, the Tourette was a new departure for Progress which was, however, hardly much larger than a scooter. The Villiers 197 cc 7.6 bhp engine sat just in front of the single rear wheel. The 8ft 6in (259 cm) car weighed only 430 lbs (195 kg) and was claimed to reach a hair-raising top speed of 55 mph (88 kh/h). Soft and hard-tops were available to cover the occupants — the manufacturers optimistically stated that *three* adults could fit in the cockpit.

However, only a handful of Tourettes were built during 1957.

The Tourette Supreme, seen here with the rare hard-top which hinged forward for entry.

SCOOTACAR

The Scootacar was the only British attempt to produce a true 'bubble' car, as inspired by the Isetta and Heinkel, and not just a microcar with three wheels. Even then, it came out looking decidedly odd, with an improbably tall (5ft) body which looked like it *had* to topple over on corners.

In fact, the Scootacar worked remarkably well in practice. The driver sat on the front end of a narrow tandem seat. A single passenger could sit

A more curious sight on the roads there could hardly have been in 1958: the Scootacar Mk I.

astride the engine itself, rather like a motorcycle pillion, or two very friendly types could just squeeze in either side of it. Steering was by handlebars, which gave a very direct response and handling was better than might be imagined, as most of the weight did in fact sit quite low down.

The origins of the Scootacar, it is rumoured, began when the wife of the director of Hunslets, the railway locomotive makers based in Leeds, said she wanted something easier to park than her Jaguar. The Scootacar was certainly that: its length was only 7ft 7ins (231 cm) and the whole thing weighed just over 500 lbs (230 kg), thanks to its glassfibre bodywork.

The first Scootacars, made from 1958, came equipped with a Villiers 9E 197 cc single-cylinder two-stroke, which produced all of 8 bhp, delivering it by chain to the single rear wheel. Being fairly streamlined for this type of car, that was enough to power it up to a claimed 51 mph.

In 1960 came the Mk 2 De Luxe version which looked rather different, with a more bulbous front and an elongated tail. The seating arrangement was different, too: the driver now had a more comfortable individual seat and the rear passengers no longer needed to apologise for digging each other in the ribs: there was enough elbow room for all.

The twin-cylinder 16 bhp 324 cc Scootacar Mk 3 De Luxe Twin arrived in 1961, selling for a fairly hefty £50 premium over the Mk 2 version.

Perhaps it was this, combined with what must have been the sheer terror of driving one at its top speed of 68 mph, which kept it from being popular. Of the 1,000 or so Scootacars built in total, a mere 20 or so were the more powerful De Luxe Twin variety. Hunslets put a stop to Scootacar production in 1965.

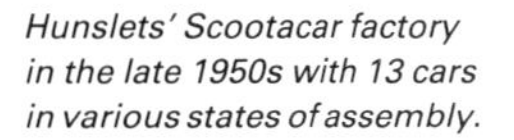

Hunslets' Scootacar factory in the late 1950s with 13 cars in various states of assembly.

CORONET

The Coronet represented an attempt to offer full-size car refinements in a microcar package. It was an open two-to-three-seater with a single rear wheel, in front of which its 328 cc Excelsior twin-cylinder 18 bhp engine was fitted. The bodywork was made in glassfibre by the coachbuilders, James Whitson & Co, who also made and assembled most of the rest of the car. Its chassis employed suspension and steering components from the Standard 8.

When it appeared in 1957, it was advertised as "the world's best three-wheeler." It was capable of a top speed of 57 mph and did at least offer big car looks. But by 1960, the Coronet was out of production. Probably around 250 had been made.

A glassfibre body and large-car looks for the Coronet.

NOBEL

Like the Isetta and Heinkel, the Fuldamobil (see page 46) was another German microcar which found its way into production in many countries worldwide, probably more than any other micro.

It also came to Britain thanks to York-Nobel Industries whose proprietor, Cyril Lord, gave financial weight to the project. With a touch of razamatazz, Soraya (the ex-Shah of Persia) was hired as the director of the works. The model produced under licence in Bristol and Newtownards, Northern Ireland was the Fuldamobil S-7, introduced in 1958. It did differ in one important respect: the German car usually had four wheels whereas the Nobel 200 was also, and more commonly, available with only three.

In other respects, it was essentially a Fuldamobil: a 191 cc Fichtel & Sachs single-cylinder engine, and GRP-and-plywood bodywork atop a steel tube chassis. Cars could be bought either fully-built or in kit form.

Ambitious plans included a weekly production rate of no less than 400 cars but it quickly became obvious that the demand for the Nobel was nowhere near that sort of figure. There were even reports that a batch of Nobels were buried under the A12 in Essex and there were certainly still Nobels in dealers' showrooms two years after the demise of the firm in 1962. Moves to merge York-Nobel with Lea-Francis, itself in the process of winding up, sealed the Nobel's fate. In all, about 1,000 Nobels were made.

Pictured in front of the Atomium in Brussels, the Nobel 200 and the open-topped Nobletta.

There were also several publicity shots of an open 'beach car' Nobel, dubbed the Nobletta, although whether Nobel or Fuldamobil actually made these is unclear; certainly Fulda built a batch for the South African market.

GERMANY

ISETTA/BMW

A fridge and motor scooter manufacturer might seem an unlikely candidate to begin the bubble car craze, but that was precisely what Renzo Rivolta did. Ironically, he never cashed in on the boom — that was left to the German licensee, BMW. Rivolta ran a firm called Isotherm in Milan, Italy, and decided to branch out into car manufacture in 1953. His Iso Isetta was the very first 'bubble' car.

The most striking feature of the new Isetta was its curious egg shape. Such forms had been seen before (the purest 'bubble' of all was the French Arzens L'Oeuf — or 'Egg' — of 1942), but it was a very well exe-

The Italian-built Iso Isetta lasted just a couple of years; note the low-set headlamps and cooling 'gills'.

cuted design. It featured a single swinging front door which allowed access to the two passengers. To ease entry, the steering column also swung out with the door on a universal joint. A folding roof was standard, just in case there should be an accident and the passengers trapped inside.

Another unusual feature was the very narrow rear track: just 20 inches across. This obviated the need for a differential — a good example of the microcar's quest for simplicity. The engine was a 236 cc 9.5 bhp two-stroke twin-cylinder unit which was placed ahead of the rear right-hand wheel to counterbalance the weight of the driver. It could power the Isetta to a top speed of 52 mph.

The reputation of the new bubble car was boosted by its unlikely participation in the Mille Miglia races of 1954 and 1955, in which all cars completed the 1,000-mile course, taking first place in their category. There were convertible, pick-up and van versions of the car, the latter known as the Iso-Carro. A larger Michelotti-styled 400 cc estate car, dubbed the Parad-Iso, never proceeded beyond the prototype stage.

Italian drivers were not drawn to the Isetta. They were well served by the Fiat 500 Topolino and, later, the Fiat 600, and did not buy the Isetta. This is why Rivolta abandoned production in Italy in 1955, having made around 6,000 examples. When he returned to car manufacture in 1962, it was with the diametrically opposed Iso Rivolta, a thundering Corvette-powered GT.

But he sold the rights to produce the Isetta to various countries: one of the the most popular was the French version, where it was made as the Velam-Isetta from 1955. It undercut the price of Citroën's 2CV by 15% at 297,000F. Its engine was the Iso 236cc unit, but the bodywork was substantially redesigned and there was a convertible version. To combat falling sales, Velam tried a luxury(!) version called the Ecrin in 1957, but by then it was too late: Velam went down the same year, having made 7,115 cars. In Brazil, the Isetta cropped up as the Romi-Isetta (3,090 examples); and it was also built in Spain.

BMW's licence to produce the Isetta began in 1955. The resulting bubble took off all over Europe.

A BMW-Isetta sits alongside the expanded BMW 600, which had a larger engine, a side door and room for five adults.

1957 BMW-Isetta had revised bodywork incorporating sliding glass instead of the previous three-piece arrangement.

But it was the German licence which led to the most familiar and successful Isetta. The ailing luxury car and motorbike maker, BMW, was literally rescued by the success of its version of the Isetta, which it began making in 1955. In place of the original engine, BMW fitted a modified version of one of its own 'bike engines, a 247 cc 12 bhp single-cylinder four-stroke unit. BMW named the new car the Motocoupe. In tests, it reached a top speed of 54 mph and returned 61 mpg.

For export markets, a 295 cc 13 bhp engine was fitted from 1956, in which guise the car was known as the BMW-Isetta 300. A three-wheeler version was also developed for export. In 1957 came a revised version with new 'one-piece' side glass which incorporated a sliding mechanism for ventilation; this replaced the three-window fixed glass inherited from the Iso.

In 1956 alone, the Isetta sold 22,543 examples — almost four times the total production of the Italian Isetta. Continental customers lapped them up and the end of the line was not reached until 1962, by which time the remarkable total of 161,728 BMW-Isettas had been built.

BMW sold a licence to produce the Isetta in Britain, where Isetta of Great Britain manufactured the model at a converted locomotive factory — where there was no road access, all deliveries had to be moved by train! From 1957 to 1964, Isettas were built in three- and four-wheeled forms, with saloon, convertible and pick-up bodies, at a rate of up to 175 per week. Estimates put the total quantity built at between 20,000 and 30,000.

The French licensed version of the Isetta was called the VELAM and had extensively revised bodywork, including a convertible option.

In Germany, there was another significant development: the BMW 600 of 1957. This was an expanded version of the Isetta, intended to offer seating for four and improved performance and refinement. Although it retained its front-opening door and family shape, it was an entirely bigger car which added rear seats (accessed by an extra side door). The engine was an in-line two-cylinder 585 cc unit developing a heady 19.5 bhp. The 600 could reach a top speed of 63 mph and was much better suited to longer journeys. It also marked BMW's first use of semi-trailing arm rear suspension.

Although the 600 only lasted until 1959, it sold very well: in total,

34,813 were made. Licensed production also occurred in Argentina under the name De Carlo during 1960. The 600's short production life betrays the fact that the opposition was getting better: the 600 was no match for the 'proper' small cars against which it competed on price.

So BMW forged ahead with the 600-based 700 Coupé, styled by Michelotti, from 1960. A saloon and a convertible soon followed. The original 697 cc engine developed 30 bhp and allowed a top speed of 75 mph; the later Sport engine squeezed out an extra 10bhp for 82 mph. By the time 700 production ended in 1965, a total of 187,821 had been made.

It is safe to say that these models sustained BMW and were responsible for financing the development of the 1500 and 2000 models which established BMW as a maker of quality saloons. The tally of nearly 400,000 microcars proves it, however uneasily this may sit with the current-day fanatics of M-Technik BMWs.

The 'other' bubble car was the Heinkel, built without the folding steering column of the Isetta and with 2+2 seating.

HEINKEL

The Isetta was undoubtedly the inspiration for Ernst Heinkel to launch his own bubble car. He was the man behind the design of the Saab three-cylinder two-stroke engine and had developed a single-cylinder four-stroke for use in a motor scooter, which proved highly successful.

He used this same 175 cc unit in the Heinkel Kabine Cruiser of 1956. Outwardly it resembled the Isetta (initially there was consternation from BMW), but that was superficial. The egg shape and opening front door were familiar, but the Heinkel did not use a folding steering wheel and was lighter and prettier. It also had more space inside, even offering children's seats in the rear. And it was produced as a three-wheeler (although a 204 cc four-wheeler was also made).

Ernst Heinkel died in 1958, just after his little car was withdrawn from production in Germany with a little under 12,000 cars sold.

Like the Isetta, production continued for many years in other coun-

This Heinkel modified by Michael Shepherd looks very much like the convertible prototype developed in Ireland.

Trojan built the Heinkel in Croydon and even essayed this delivery van version.

tries. In Argentina, the local Heinkel was built from 1957 to 1959, and 2000 examples were sold. In Ireland, where production transferred from Germany in 1958, the model was known as the Heinkel-I. Somewhere in the region of 8,000 cars are estimated to have been built in Ireland until 1961, when the licence transferred to Trojans in the United Kingdom. The Irish firm had developed a full convertible version called the Open Tourer, although it was sadly not productionised.

The Croydon-based firm Trojan had made a number of vehicles since 1922, although production had ceased before the Second World War. Its decision to take on the Heinkel marked its return to car manufacture. It made cars in left- and right-hand-drive form, from 1961, latterly with the name Trojan 200. Trojan also made an Estate Van version towards the end of the model's life. By 1965 it had become disenchanted with car production (including the Elva sports car range which it had also taken on) and the bubble was abandoned, after some 7,000 cars had been made.

FEND/MESSERSCHMITT

The name Messerschmitt is at least as familiar for its cars as for its aeroplanes. Following the war, an engineer with Messerschmitt, Fritz Fend, applied himself to peace-time manufacturing and initially designed pedal-powered transport and vehicles for veterans injured in the war.

The first road car was the Fend Flitzer of 1948, a tiny single-seater with the most basic specification: bicycle wheels and, in the first versions, pedal power. Those unable or unwilling to use their legs could opt for what was one of the smallest engines ever fitted to a motor car: a 38 cc single-cylinder Victoria unit — powerful enough to take the 165 lbs car to a heady top speed of 19 mph. Unhurried drivers were rewarded by a claimed fuel consumption of 235 mpg. Both enclosed and convertible versions were offered, the latter with a transparent inflatable hood! 30 examples were made before its replacement arrived in 1949.

The next Flitzer used motor-scooter wheels and a 98 cc Fichtel & Sachs engine developing 2.5 bhp, giving a top speed of 38 mph. 98 of this type were made.

In 1950 came the improved Kabinenroller, a marginally more substantial open or closed single-seater with lines even more portentous of the Daleks of *Dr Who*. It used a more robust 98 cc Riedel engine (4.5 bhp, 47 mph). Fuel consumption was down to a 'mere' 100 mpg. Fend made 154 before he began his famous collaboration with Professor Willy Messerschmitt in January 1952.

Messerschmitt had not been allowed to build any more aeroplanes after the end of the war and had devoted his works to the repair of railway rolling stock. He was an admirer of Fend's ideas and decided to go ahead with a small car project with him. The first fruit of this get-together was the Rikscha, a bizarre 125cc three-wheeler where the driver sat as on a scooter behind a sort of scoop which could carry two passengers. It did not enter production.

More serious — and influential — was the first Kabinenroller. The first models bore Fend's own name and were sold as the Fend FK 150 from early 1953. They had 148 cc Fichtel & Sachs 6.5 bhp engines mounted in the rear.

Shortly after, in March 1953, Messerschmitt announced that it would be selling the model under its own name with a larger 174 cc Fichtel & Sachs engine, which established the classic form of all production Messerschmitts. It had many striking features, but the most commented upon was the Plexiglass canopy which resembled so closely the canopies of Messerschmitt's wartime aircraft. The whole canopy swung up sideways to allow entry for two passengers, who sat in tandem in the narrow body. A pair of bug-like headlamps provided illumination and the front wheels were enclosed in hemi-spherical wings. If science fiction had been a popular form in 1953, the Messerschmitt would have been greeted as an alien pod.

A line-up of Fritz Fend's cars: a Fend Flitzer makes an interesting comparison with later Messerschmitts.

An early Messerschmitt KR175, identifiable by its faired-in front wheels which made turning tight corners difficult.

The factory made great play of the KR200's ability to climb mountains, as proven by numerous crossings of difficult Alpine passes.

The KR201 was the convertible version of the 'Schmitt which had the advantage of not frying the occupants' heads in strong sunlight!

The production KR 175, as the car was known, had a 174 cc Fichtel & Sachs single-cylinder two-stroke engine sited in its rump which pumped out 9 bhp. Even with such little power, the lightness (460 lbs) and aerodynamics of the 'Schmitt gave it a top speed of 56 mph and established it as one of the 'performers' of the microcar scene. It was steered by a pair of handlebars which incorporated a twist-grip throttle. By the time it was replaced in 1955, 19,668 had been built.

Its replacement was the KR 200, whose engine had grown, as the name suggested, to 191 cc and 10 bhp. The body was slightly different, too: there were cut-outs in the front wings to allow the front wheels to steer more freely and so give something better than the appalling turning circle of the KR 175. The floor-mounted accelerator was new and the 'Schmitt was also now able to go in reverse — by restarting the engine in the opposite direction, allowing the hair-raising possibility of travel in reverse in four gears!

There were better interior appointments, including some rather unusual options. One show car appeared with mock-alligator skin upholstery, there was the option of a pretty chunky valve radio and you could even order a set of mounts for skis!

The KR 200 was a huge success: more were sold in 1955 alone than all previous KR 175 production. But there was now severe competition from

Fritz Fend pilots his Tg500 'Tiger', which was a genuine sports car and the only four-wheeled Messerschmitt.

other marques such as BMW, and Fend organised some publicity stunts, like the antics of the high-speed streamlined KR 200 Super, which broke 25 class speed records in 1955 (top speed: 87 mph).

On the commercial front, the appeal of the model was significantly extended by the arrival of a cabriolet version in 1956, dubbed the KR 201.

As a result of court action by Mercedes Benz, who obscurely claimed that Messerschmitt's 'flying bird' symbol looked too much like its own three-pointed star, the FMR triple diamond badge appeared on Messerschmitts from January 1957: this stood for Fahrzeug und Maschienenbau GmbH Regensburg.

Later that year came what was undoubtedly the zenith of all microcars: the Tg500, or Tiger. This was perhaps more properly viewed as a cheap sports car than a microcar as such, for it had four wheels and, in microcar terms, a stonking engine fitted in the form of a 493 cc Sachs. It developed 20 bhp and Messerschmitt claimed that the Tg500 could exceed 85 mph, although a contemporary road test achieved only 68 mph and 0-60 mph in 27.8 seconds. Production of the dome-top Tg500 began in 1958 and the Roadster version came the following year. But initial enthusiasm was quenched by customers' increasingly sophisticated tastes. Probably only about 450 were made.

There were a couple of other aborted Messerschmitt prototypes. In 1958, FMR built a much more conventional-looking four-wheeled saloon which, it stated, was to be built in the United States under licence with a 400 cc or 600 cc rear-mounted engine. Nothing ever came of the plan. And in 1962 an intended successor to the KR 200 was built, with a longer snout, separate front wings and quite fantastically wild rear fins. The steadily decreasing popularity of the KR put paid to that car.

Without doubt, the Messerschmitt caught the bubble boom in full swing and was one of the more popular and attractive designs. It had its faults, like skittery road behaviour and lack of refinement but it was better than most microcars by a quantum leap. Almost from the beginning, Messerschmitts had been making a loss for the parent company, which survived by making bottle top dispensers and continued the microcar rather as a labour of love.

The very last examples of the KR and Tg500 left the Regensburg works in 1964, by which time over 30,000 KR 200s had been supplied. Apart from an abortive attempt at licensed production in Italy in 1954 (where the car was to be known as the Mivalino), that was that for the Messerschmitt.

In 1990, a German firm offered the Leonhardt Tiger, a vague replica of the Messerschmitt, fitted with a tuned Mini engine.

LLOYD

The origins of the Lloyd name go back to 1908, but Lloyd as a firm was quickly subsumed into the Hansa/Borgward combine and dropped as a marque. Until, that is, April 1950, when the combine revived the name on a new small saloon, a rare example of a true microcar being made by a large motor manufacturer.

The Lloyd 300 was designed to accommodate four people (which it did, but only just) to sell for just DM 2500. It had a 293 cc air-cooled two-stroke twin developing 10 bhp mounted in the front.

While it succeeded in looking like a 'real' car, the Lloyd was very simply constructed: a steel chassis platform with a wooden frame to carry the fabric body panels (a material chosen because steel was in such short supply).

Although its launch price was rather higher than expected (at DM 3334), it was still over 25% cheaper than a VW Beetle and sold strongly: it took under two years to sell 10,000 cars. These included many estate cars (which bore the name KS 300), but the two-seater coupé (the LC 300) was a commercial failure.

The replacement Lloyd LP 400 and Kombi LS 400 arrived in January 1953, with 386 cc 13 bhp engines. The 400 quickly did away with the fabric construction and subsituted it with (now more freely available) steel. A 250 cc version — naturally called the LP 250) was also sold from 1956 (the justification for which was that it could be driven on a motorbike licence) but it lasted only one year and sold a mere 3768 examples. The last LP 400 was made in 1957.

Alongside the LP 400 came the LP 600 in 1955. This had an all-new 596 cc four-stroke two-cylinder engine which developed 19 bhp and was joined in 1957 by a more powerful version (dubbed the Alexander). The Alexander TS which followed in 1958 had 25 bhp on tap and could out-accelerate a Beetle, although it now cost over DM 500 more than the VW. All three models were available until 1962.

There was even a 'microbus' based on the 600. Called the LT 600, it was a six-seater three-door estate car with a quite different appearance to the other Lloyds.

As Lloyd faced declining sales for its cars, it moved towards larger engined cars with the 900 cc Arabella of 1959, and there was only one more 'micro', the heavily stylised Alexander Frua Coupé, marketed from May 1959. It used the 25 bhp 596 cc engine in a Pietro Frua-designed Renault Floride-like body and had utterly anaemic performance for all the promise of its looks. Only 49 were built.

As the Borgward combine hit financial problems, the Lloyd name sank too, on 4 August 1962. A remarkable total of over 300,000 Lloyd microcars had been made by that time.

Lloyd's LP400 was a successful attempt to marry microcar economies with large-car qualities.

The Lloyd Alexander Frua was an extraordinary coupé using the front-mounted 25 bhp Lloyd engine.

The Lloyd LP600 (right) and Alexander TS Kombi (left).

KLEINSCHNITTGER

Paul Kleinschnittger was a keen motor engineer based in Ladelund, Germany. His passion for small cars was realised in his F125 of 1950, developed and productionised in just nine months.

Kleinschnittgers were immediately popular. Although they looked like toy cars, they were well-designed machines with good handling and fair performance (44 mph or 70 km/h) from a front-mounted Ilo 125 cc single-cylinder engine, which also drove the front wheels. The F125 featured rubber suspension and a gearless differential. Its specification was also unusually complete. At a cheap price of DM 1995, production began at the encouraging rate of 50 cars per month.

The F125 weighed only 330 lbs (150 kg), which was helpful when Kleinschnittger decided to run a works racing team — which scored a second placing behind a Porsche 356 in the under-1100 cc class in the Lisbon-Madrid rally — not bad for a 125 cc engined car!

Setting his sights on the then more popular larger-engined micros, Kleinschnittger developed the F250 in 1954. This was a metal-bodied coupé with a front-mounted 246 cc Ilo engine but it never truly went into production.

Despite its toy-like appearance, the Kleinschnittger offered many virtues.

The following year, the coupé reappeared alongside an alternative coupé, called the F250 S (the original model was renamed the F250 C). This was an unusual three-seater coupé (the passengers sat either side of the driver!) which shared the same basic specification and bodyshell as the C. There was also a two-seater coupé/cabriolet called the F250 Super.

In fact, it took until 1957 to get the new F250 series into production and it was only the F250 Super which was made in any numbers (22 were built). The financial strain of developing the F250 models forced Kleinschnittger to close on 5 August 1957. According to records, by this time, the remarkable total of 2,980 examples of the smaller F125 had been made.

Dornier's original symmetrical car of 1955 was the Delta.

The Delta was developed by Zundapp to become the Janus, which had opposite facing rows of seats and a central engine.

ZUNDAPP

Zundapp was one of Germany's best known motorbike manufacturers before it took over the rights to the strange little Dornier Delta in 1955. The Delta had been shown at the Frankfurt Motor Show that year and featured a curious system of twin doors, one at each end of the car, which hinged upwards to allow entry to two sets of two passengers, who sat back-to-back. It was a completely symmetrical car.

Zundapp developed the idea into the Janus, which was not productionised until 1957, by which time the BMW-Isetta and Goggomobil had a firm hold on the market. The name Janus echoed the Roman god who faced two ways. The doors were changed so that they opened sideways but the shape remained broadly symmetrical.

The Janus' engine was a 248 cc single-cylinder two-stroke unit developing just 14 bhp, sited between the front and rear seats. A top speed of 50 mph was claimed, which was not even a match for the Isetta. As the Janus cost significantly more, Zundapp's sales target of 15,000 a year was drastically optimistic. Only 1,731 were sold in 1957, followed by 5,169 the next year, when Zundapp was forced to pull the plug on its car project: a total of only 6,800 had been made. Zundapp fell back to concentrating on its motorbikes.

Dornier made a few other prototypes in the late 1960s and early 1970s: this is the Delta II of 1969.

TRIPPEL

'Schwimmwagen', or amphibious vehicles, were Hanns Trippel's speciality: during the Second World War he built about 1,000 Trippel SG6 amphibians in Bugatti's Molsheim works. After the war, in 1949, he made a miniature aquatic version, dubbed the SK 9, which used a 300 cc DKW engine in the tail, but that remained a prototype.

Trippel began work on a true landlubber, an enclosed coupé to be called the Trippel SK 10, in 1950. This was one of the more sophisticated microcars of the period, thanks to the experience of its creator. It was a three-seater using a 498 cc Zundapp two-cylinder engine in the tail. In standard form, this developed 18.5 bhp — enough to take the 1,056 lb (480 kg) car to a top speed of 72 mph (115 km/h). In the sports version (26 bhp), the SK 10 could reach a remarkable 88 mph (140 km/h).

In appearance, the SK 10 was remarkably mature, rather resembling a Porsche 356. The Laupheim-based coachbuilders, Bobel, also made a convertible version, but that did not truly enter production. Trippel had abandoned production of the SK 10 by 1952.

The convertible (and the coupé) were made under licence in France by the firm SIOP (which also produced the Rosengart) from 1953. The open car was known as the Marathon Pirate, the latter as the Marathon Corsaire. These had slightly modified glassfibre bodywork and Panhard Dyna 746 cc engines producing up to 52 bhp — sporting indeed. The Marathon eventually fell victim to SIOP's financial crisis in 1954. Few had been built, although licensed production also took place in Holland and Austria.

Back to Hanns Trippel, now based in Stuttgart. He decided that, despite the setbacks, he could still make a go of his undeniably good design. Expanding the essential structure of the Corsaire, he built the Trippel 750 with a 750 cc Heinkel three-cylinder engine and seating for 2+2. Both open and coupé bodies were built in 1956. A Belgian firm, Wilford, was to produce the new car, but it too suffered financial strife and the 750 was left to another firm to revive. The German firm Weidner modified the design again and sold it as the Condor from 1957-58. But it was rather too expensive for such a small car and Weidner had to abandon production having made about 200 cars.

Trippel went on to design the world's most successful amphibious car, the Amphicar, in 1959.

Hans Trippel's SK10 was sometimes dsecribed as a mini-Porsche; its 498 cc engine sat in the tail.

This convertible version of the Trippel SK10 was built by Bobel and shown at the 1952 Geneva Motor Show.

FULDAMOBIL

If the first Fuldamobil of 1950 looked like a three-wheeled caravan, it should come as no surprise: its builder Norbert Stevenson relied heavily on caravan construction techniques to build his new microcar. Hence, it had a wooden skeleton with steel panels simply nailed to it.

Production of the Fuldamobil three-wheeler began in 1951 with a slightly improved construction and appearance: a wooden frame was now panelled with leather-faced plywood sheets. A single-cylinder Baker & Polling 248 cc engine sat in the back, bolted directly to the floor; one can imagine the shaking that went on for the 400 or so buyers of these early cars.

With the N-2 of 1952, matters improved rather. The more rounded bodywork was now in unpainted hammered aluminium panels, which earned it the nickname of 'silver flea'. Like the first Fuldamobil, it was sold in coupé and roadster forms, but the former was now a 2+2 seater, rather than a strict two-seater. It was also slightly lighter. This, combined with the extra power of the Fichtel & Sachs 359 cc single-cylinder engine, gave it a more useful turn of speed and a fuel consumption of over 60 mpg.

The N-2 lasted three years to 1955, by which time it had sold a modest 380 examples. As the 'bubble boom' was now taking off in Germany, the Fuldamobil really missed the boat as far as sales were concerned.

In 1953, Stevenson built a prototype with a new, more rounded body, incorporating a hinged bootlid in its aluminium bodywork. However, his own firm Elektromaschinenbau Fulda, did not put it into production: instead, a licence was sold to the bus manufacturer Nordwestdeutsche Farzeugbau (NWF), which built the so-called 'S' series under the name NWF 200 in the years 1954-55. The engine was the ubiquitous Ilo 197 cc single-cylinder, which produced just the same power as the old 359 cc engine (9.5 bhp).

This engine was favoured by Fulda in its own version of the 'S' model, which it named the S-2. It sold for a small premium over NWF's version, with which it was concurrently sold. About 430 were built to August 1955, as compared with over 700 of the NWF type. All were saloons: Fulda cabriolets were abandoned from this time (with the exception of an S-5 prototype in 1956).

The S-3 was a 191 cc Fichtel & Sachs engined prototype, the production version of which was called the S-4, introduced in late 1955. It marked Fulda's shift away from three-wheelers, favouring the close-set twin rear wheel system adopted by Heinkel and BMW with the Isetta; a version with only one rear wheel remained available to order. With the popular smaller engine fitted, the S-4 sold for considerably less than the outgoing S-2: only DM 2780 versus DM 3350. By the time it and its slightly improved successor, the S-6, left production in 1957, some 300 examples had been sold.

1957 was the year of revolution for the Fulda. With the S-7 of that year, not only was there a revised body style, the body material changed too, to glassfibre. Among the advantages of glassfibre were its lighter weight, which gave it improved performance and economy from the same 191 cc engine.

But by then, Fulda had already missed the big bubble band wagon and declining sales set in. Not that this really affected its availability: the Fuldamobil was probably the last surviving German microcar firm of the 1950s, still offering the S-7 as late as 1969, by which time about 700 had been sold. From 1965, these had all been fitted with 198 cc Heinkel engines.

The Fulda legend lived on across the world, however, with licensed production being taken up in numerous countries. The most famous was in Britain, where York Nobel built his Nobel 200 (see page 31). The rickshaw-style Nobletta was also built by Fulda as the Sporty, most of which were exported to South Africa. There was also licensed production in Holland (where the car was known as the Bambino), Sweden (Fram King Fulda), Norway, Chile and Argentina (Bambi), India (Hans Vahaar) and Greece (Attica, then Alta).

Early Fuldamobil N-2 had a very basic construction and an engine bolted directly to the bodywork.

The later Fuldamobils featured more rounded glassfibre bodywork and more sophisticated construction.

GLAS GOGGOMOBIL

Hans Glas rose to become a legendary figure in the German motor car industry. His post-war move from agricultural machinery to scooters, microcars and, ultimately, mainstream sports cars and saloons was dramatic.

After making scooters at his Dingolfing, Bavaria factory for a coupé of years, Glas decided to create a small four-seater car in 1953. The first prototype had a front-opening door like the Isetta but by the time the production version appeared in 1955, the Goggomobil sported conventional side doors. Its shape was unmistakable and introduced large car features into a very compact (9ft 6ins/290 cm long) vehicle.

The Goggomobil T250 was powered by a rear-mounted two-cylinder two-stroke engine of Glas's own design. As the car weighed only 770 lbs (350 kg), performance from the 14 bhp unit was sprightly: 50 mph (80 km/h) and 50 mpg were quoted. Four adults could easily sit in the cabin and the response from the public was immense: within six months of launching the T250, 5,000 had been sold. By the time the T250 left production in late 1956, almost 25,000 had been made.

The T300 and T400 then became Glas's standard Goggomobils. These had larger and more powerful engines (up to 20 bhp) and a better specification. In Britain the Goggomobil was known as the Regent and its new sister coupé model of 1957 was called the Mayfair.

The coupé (known as the TS300 or TS400 depending on engine size) was a pretty little thing sharing the same floorpan as the saloon. It was somewhat heavier (970 lbs or 440 kg) and had only two seats but was faster at 65 mph (105 km/h).

In March 1958, the 100,000th Goggomobil was made, making it easily the most popular microcar of the day (and one of the most popular of all time). This reflected its huge successes both in Germany and export markets. People even campaigned them in motorsport with some success.

Also in 1958 came the Goggo's 'big brothers', the T600 and T700, sometimes referred to as the Isar. Much larger than the little Goggo, at 11ft 3ins (343 cm) long, the cars' big new flat twin four-stroke engines gave more power (up to 30 bhp), but were unreliable. Nevertheless, up to 1964, some 87,575 of these larger cars were built, including estate versions.

The Glas Goggomobil was one of the best-selling microcars of all time and continued in production until 1969.

In coupé form, the Goggomobil TS300 could achieve a top speed of 62 mph.

Several projects never saw production: there was a Goggomobil Coupé (the S-35) with a 700 Isar engine in the front in 1959; and the 'M61' prototype, a 400 cc coupé designed to fill the gap between the Goggo and Isar.

In Australia, the Glas agent, Bill Buckle, fitted his own glassfibre copies of Goggo bodies on imported Glas chassis and even made a glassfibre sports shell called the Dart using a 400 cc Goggomobil engine from 1959. And in Spain, licensed production was carried out under the name Munguia.

In 1964, the door hinges of the Goggomobil were moved to the front pillar. As most other microcars and bubbles had burst by this time, the Goggo was able to continue in production but even its sales were rapidly declining. BMW eventually took over Glas's operation in 1967 which led to most of Glas's larger cars (which had been in production since 1961) being pensioned off; only a few remained as badge-engineered BMW models.

However, the Goggomobil outlasted them all. BMW finally pulled the plug on it in 1969. By this time, some 280,000 cars had been made — a remarkable total for a microcar. The Goggomobil just outlived its creator, as Hans Glas died in 1968 at the age of 78.

There was even a follow-up in 1971: a Berlin firm introduced a Goggomobil-based car called the AWS Piccolo (from 1974 known as the AWS Shopper), which lasted until 1976.

A stillborn attempt at a larger Goggo coupé: the Glas Isar S-35 prototype of 1959.

BRÜTSCH

Among the microcar world's prolific engineers, Egon Brütsch ranks right alongside Lawrie Bond as the creator of the most designs. But despite his unstoppable quest to create the perfect microcar, he hardly sold any cars at all. His best-selling model, the Mopetta, scored just 14 sales. In truth, Brütsch was more of an inventor and publicist than a manufacturer.

Brütsch came from a wealthy background and initially made single-seater prototypes for children, then for adults. The first cars were crude machines which were not really practical for a production run (although a Lloyd 400 cc engined single-seater was to have been made in the Netherlands). After experimenting with larger sports bodies, Brütsch was inspired to create a new microcar using a glassfibre body.

So he built the Brütsch 200 Spatz, which was presented at the 1954 Paris Salon. It was a smoothly shaped, little, open three-wheeler which looked sporting and attractive. However, it had no chassis, relying solely on the strength of the plastic shell to keep the mechanicals in place. This was in the early days of glassfibre when the strengths of the material were often over-estimated: the 200 proved to be a very weak structure and cracked in many places.

The German company which bought the rights had to redesign the car substantially and the result was the Spatz ('sparrow'), shown in 1955 (see page 52). Licensed production of the original 200 was carried out in Switzerland under the name Belcar, but was a short-lived project.

Brütsch then built the Zwerg ('dwarf') in 1955, which was made in one- and two-seater forms, the former with a 75 cc engine, the latter with a 250 cc Maico unit. About 12 were made and more were built in France under the name Avolette.

The next project was the Mopetta of 1956. This car can justifiably challenge the Peel P50 as the smallest car ever built. It was only 67 inches (170 cm) long, 34.7 inches (88 cm) wide and 39 inches (100 cm) high. It weighed just 134 lbs (61 kg) and had a 49 cc engine developing 2.6 bhp, enough to give it a top speed of 27 mph (45 km/h). In appearance it resembled an egg in the process of hatching its driver.

The original publicity stated that the Mopetta could float in water. Nobody ever tried, which is perhaps just as well, since the engine sat exposed beside one of the rear wheels! Georg von Opel (late of Opel) was all set to produce the car in series as the Opelit, but the project came to nought, mainly because the authorities refused to allow it on the road. Most Mopettas were sold in Britain.

Egon Brütsch's first serious microcar prototype, the 200 Spatz, never made production as it was declared unsafe for the road.

Brütsch's Zwerg (dwarf) was a development of the 200; here it has its optional and very unusual plastic canopy.

A slightly larger sister, the Rollera, arrived at the same time, offering 98 cc and a top speed of 50 mph (80 km/h); again, there was licensed production in France.

Also in 1956, Brütsch developed the Bussard ('buzzard') and Pfeil ('arrow'), which were three- and four-wheeled versions of the same basic bodyshell, the former with a 191 cc engine, the latter with a 386 cc unit. Very few of each were made, although both models were sold alongside the Mopetta in Britain.

The inveterate microcar creator then moved on to the V-2 (1957), which was an attractive two-seater with a choice of 98 cc or 247 cc engines. Twelve examples were made.

Brutsch's last laugh was the V2N of 1958. This was based on the V-2 but had a Fiat 500 engine in the tail. Its polyester body had more than a hint of American fins-'n'-chrome about it, although dimensionally it was no more than 124 inches (315 cm) long. The 'N' in V2N stood for Ngo, the name of the Indonesian who built Brütsch cars in his own country and wanted to start production of the V2N in France. This French version, the so-called V2N Jet, was quite drastically restyled (not very attractively) and was to be sold with 125, 175 and 200 cc engines, but production never began.

The driver of the miniscule Mopetta looked like he was being hatched by the car; it was also claimed to be amphibious!

Development of the 200 Spatz: the Bussard three-wheeler arrived in 1956.

The four-wheeled version of the Bussard was called the Pfeil ('Arrow') and, like its brother, was sold in Britain.

Brütsch's Rollera was a larger version of the Mopetta and was made under licence in France (where this example was built).

The V2N was the last of the long line of Brütsch cars: it had American styling traits and sported a Fiat 500 engine in the tail.

SPATZ

Having bought the licence to produce Brütsch's 200 Spatz, the German machine maker Alzmetall was faced with a car which could never go into production. A court ruled as much when Egon Brütsch took legal action over Alzmetall's failure to pay him licence fees: the judge stated that the 200 would have been unroadworthy because of its all-GRP construction.

The reason Alzmetall did not pay any fees was because, through the hands of 77-year-old Tatra engineer Dr Hans Ledwinka, it had altered the 200 so much that it was now a new model, now called simply Spatz ('sparrow'). It debuted in 1955 and featured four wheels instead of three, a steel chassis and a much revised appearance, including headlights positioned on embryonic front wings. The engine remained the Fichtel & Sachs 191 cc single-cylinder (10.2 bhp).

Production of the Spatz began in 1956, but only in door-less cabriolet form. Spatz displayed a gullwing hardtop version but a patent then existed on the gullwing system in the name of none other than Hans Trippel, who prevented the car being sold. Almost immediately, Spatz looked for a production partner, which it found in the motor-bike maker Victoria. Together they formed Bayerische Auto-Gesellschaft (BAG).

Improvements soon followed. A 250 cc 14 bhp engine pushed the top speed up from 50 mph (80 km/h) to 62 mph (100 km/h) and the cars then carried the name Victoria 250. The last models bore the name Burgfalke 250 after the aircraft company which became involved in 1957, but by then the writing was on the wall. About 1600 cars of this type were made in total.

Several German firms produced the Ledwinka-modified Spatz, a four-wheeler based on the Brütsch prototype.

In hardtop coupé form, the Spatz had gullwing doors for entry; however, this version did not enter series production.

FRANCE

MATHIS

Emile Mathis was one of the pioneers of aerodynamic design in the 1930s and his first project following the war was a wind-cheating three-wheeler which he dubbed the Mathis VL 333. This stood for Voiture Legere followed by the trinity of fuel consumption (3 litres per 100 km – or 94 mpg), 3 seats and 3 wheels.

Mathis certainly succeeded as far as aerodynamics was concerned: the 333 had a Cd of just 0.22, thanks to the design input of an engineer called Andreau. The front of the car (where the 707 cc engine sat) was very wide at 68.5 inches (174 cm) and incorporated faired-in lights and grilles.

With its aluminium bodywork, the Mathis weighed only 970 lbs (440 kg), which gave it its strength as an economy car. It was displayed at the 1946 Paris Salon and several prototypes were made but, sadly, a production run for this advanced car never materialised. Mathis went on to create a larger vehicle with the name 666. He died in 1956.

The Mathis VL 333 of 1946 looked – and was – years ahead of its time, but never made intended production.

An insuppressible pre-war microcar creator, Charles Mochet, produced a number of post-war cars – including this CM 125 Y.

MOCHET

Charles Mochet was active before the war making small cars in Puteaux, France. In 1946 he began making *voiturettes* once again. A series of prototypes began with a pedal car and progressed through vehicles with 50 cc and 100 cc engines.

The CM Luxe of 1948 was probably the first production Mochet, and several different body styles, all of them open two-seaters, were made. The model was also referred to as the Velocar Type K.

But the definitive Mochet did not arrive until 1951, with the introduction of the CM Grand Luxe. This had the distinctive enclosed bodywork of all future Mochets and did away with the crude cycle wings of the previous cars. It had an Ydral 125 cc engine developing 3.5 bhp sited in the back.

In 1954, an improved version called the CM 125 Y was marketed, now also offered with a 175 cc Ydral engine with a much more sprightly 8 bhp to offer (top speed 44 mph or 70 km/h). This lasted until 1958, by which time around 3000 cars had been made. Mochet had also built a 750 cc convertible sports car in 1953, dubbed the CM 750, but this remained a prototype.

The Arzens Carrosse of 1951

ARZENS

Most people give the credit for being the world's first bubble car to the Iso Isetta of 1953. And indeed it was the first commercially available bubble. But the credit for creating the first bubble shape undoubtedly belongs to a French artist and train designer by the name of Paul Arzens.

Arzens decided to build an electric city car in Paris during the war, in 1942. The car's name, l'Oeuf (which means 'egg'), betrays the very simple design basis of Arzens's car. Its striking body was formed in aluminium and its curved side doors and windscreen were made in Plexiglass — a quite new technique at the time. The result was pure fantasy. Ettore Bugatti was reputed to have been most impressed when he spotted it parked outside a cafe.

Five 250 amp batteries powered the Egg via an electric motor sited on the single-wheel trailing arm. This gave it a range of 60 miles at 45 mph. The Egg never went into production but it still exists today, up until 1990 in the hands of its inventor, when he died aged 87. It is now fitted with a 125 cc petrol engine.

After the war, Arzens built another strange device named the Carrosse during 1950. His aim was to create a car which would be easy for unskilled labour to assemble. So he built a structure in steel tubes, all of which slotted on into the others, eliminating the need for nuts and bolts. It had four wheels and a 125 cc two-stroke engine, but never entered production.

Paul Arzens' silver Egg of 1942 was pictured on the cover of Life *magazine; one admirer was reportedly Ettore Bugatti.*

The pretty 1947 Rovin D2 was sold for a single year only: the catalogue described it as 'a French jewel'.

ROVIN

One of the better-known French microcars is the Rovin, the product of brothers Robert and Raoul de Rovin. Their first prototypes were made as early as 1946 and established the classic shape of the early Rovin.

The brothers installed themselves in the old Delaunay-Belleville works in St Denise, where they established a production line in 1946. The Rovin D2 sported a water-cooled 425 cc two-cylinder four-stroke engine mounted in the rear, from whose 10 bhp a maximum speed of 40 mph (64 km/h) could be expected.

It only lasted one year, during which time about 200 were built. It was replaced by the D3 in September 1948, a car which shared the mechanical basis of the old model but boasted smart all-new bodywork which was virtually symmetrical front to rear.

Robert de Rovin's D4 was quite a swanky conveyance by the standards of the day (1950).

In 1950 (after 922 of the D3 had been made), the car's engine size increased to 462 cc and the power output rose to 13 bhp. The new D4 had a top speed of 53 mph (85 km/h). It was externally distinguishable by its headlamps mounted high on the wings, as opposed to being on either side of the false front grille. The Rovin's heyday was the early 1950s, but cars continued to be offered on a much less enthusiastic basis until 1961, by which time around 1000 examples had been made.

REYONNAH

Resolving parking problems has been a constant theme of the late twentieth century. Short of cars which park on their ends (of which there have been examples), the Reyonnah surely offered the most radical solution: *une voiture reductible*.

Launched at the 1950 Paris Salon, the Reyonnah was the brainchild of Robert Hannoyer (whose name provided the anagrammatical basis for the car's name-tag). It was a narrow-bodied tandem two-seater with an Ydral 175 cc (or AMC 250 cc) engine placed in the rear, driving twin rear wheels placed only 20.5 inches (52 cm) apart.

But its distinguishing feature was the way its front wheels folded under under the bodywork. This system, devised by M. Hannoyer, consisted of a pair of parallelogram arms hinged near the body. By releasing the sponsons, the car's width shrank from 52 inches (132 cm) to just 29.5 inches (75 cm).

There were no doors on the Reyonnah, the passengers having to climb over the sides. Either a soft-top or a sideways-swinging Plexiglass dome protected the occupants from the elements. By 1954, when the project folded (if you will pardon the pun), very few examples had been built.

The extraordinary Reyonnah, born of a desire to build a car which could fold up to fit through narrow doorways.

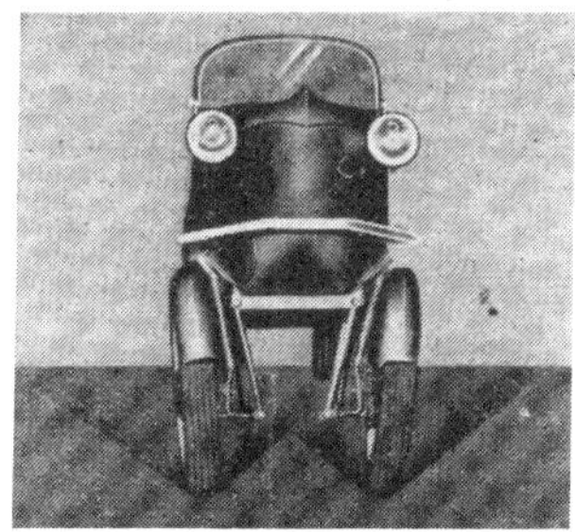

INTER

If the Messerschmitt KR175 was the aeroplane of the German microcar boom, the Inter was its equivalent in France. It even appeared in the same year, being launched at the 1953 Paris Salon and had more than its profile in common with the 'Schmitt.

Narrow bodywork and a tilting canopy were the striking features of the aircraft-inspired Inter.

Its resemblance to an aircraft was not altogether surprising since it was made by the SNCAN aircraft factory in Lyon. The two passengers gained entry by lifting the narrow dome-top sideways and sat in tandem in the very narrow body. The car (which was only 53 inches or 135 cm wide) could be made to become even narrower by 'folding' the front wheels inwards and forwards in a manner similar to the Reyonnah (see previous), so that, at just 35.4 inches (90 cm) wide, the Inter occupied no more space for parking than a motorbike.

The Inter used a 175 cc Ydral single-cylinder two-stroke engine placed in the rear and driving the single rear wheel. With 8.5 bhp on tap, this powered the scooter to a top speed of 50 mph (80 km/h).

Unlike most other French microcars of the period, the Inter actually made it into production and it is estimated that about 100 were built by the time production ceased in 1956.

VOISIN/BISCUTER

Gabriel Voisin, the creator of some of the more original French cars between 1919 and 1937, embarked on an altogether different tack after the war, with a series of strange-looking prototypes. The only one to make it into production was the Biscooter which he first displayed in 1950. But it was not in France that the car was made, but in Spain — a country which had no car industry of its own at the time.

Of the two prototypes Voisin had created (one in 1950 and one in 1952), it was the earlier and, it must be said, the more characterful one which was chosen by the Barcelona-based firm Autonacional as the one to make in series.

This was a somewhat baroque two-seater with virtually no bodywork and bare aluminium finish. The Spanish firm, in collaboration with Voisin, gave it rather more of a body and swapped the Gnome-et-Rhone 125 cc engine for a 9 bhp 197 cc Hispano Villiers single-cylinder unit, still mounted up front.

Weighing only 528 lbs (240 kg), the Biscuter (as the car was renamed for Spain) was claimed to achieve 48 mph (76 km/h) and return 57 mpg.

The first production cars were made towards the end of 1953. The basic model was an open two-seater which came to be something of a definitive form of city transport in Spain. Other body options soon appeared: a coupé with fully-enclosed glassfibre bodywork and a four-seater estate were both offered. As these appeared, the models were named the 200 R (open), 200 C (estate) and 200 F (coupé). There was even a Voisin-designed extended wheelbase open four-seater in 1955, but this remained a prototype.

With precious few other cars made in Spain, the Biscuter made a killing. Until, that is, opposition arrived in the form of the Seat 600. Nevertheless, by the time Biscuter was withdrawn from sale in 1960, about 20,000 had been sold, making it one of the most successful microcars ever made.

Gabriel Voisin's absolute-bare-minimum Biscooter prototype of 1950 was never made in its home country, France.

Instead, a Spanish firm built it – with marginally more bodywork – as the Biscuter throughout the 1950s.

OTI MICROCAR

One of the most bizarre post-war French micros was the OTI Microcar. Its first appearance was at the Gala des Artistes (a modern art exhibition) in 1957, where it was seen with open bodywork protected by Dali-esque melting rubber bumpers which circumscribed the whole car. In hardtop form, the Microcar was to have a bulging Perspex dome top! Its engine was a 125 cc Gnome-et-Rhone two-stroke.

Surrealists were to be disappointed when the production version, sometimes referred to as the Villeple, materialised in 1959. It was a rather plain contraption by comparison, lacking the bug-eye headlamps and Dr Seuss bumpers of the earlier car. One quirk did redeem it, however. It was to be built at the old Bugatti factory in Molsheim and the OTI became history's least likely recipient of a Bugatti-style horseshoe grille! Perhaps thankfully, this micro-Bugatti was never productionised.

Dali would have been proud of this 'melting' car: the amazing OTI Microcar of 1957.

VESPA

To generations of Mods in Britain, the Vespa is a legend. But the Italian scooter manufacturer which made it, Piaggio, did make a break for the car market briefly from 1957. But it was not made in Italy: it came from Nievre in France.

The reasons for this boil down to Fiat's domination of the Italian car market. There was an agreement between motorcycle makers and Fiat whereby they would not challenge Fiat's position by manufacturing *cars* in Italy. Making them abroad was OK.

The Vespa 400 was a pretty little 2+2 coupé with a rolltop, all-independent suspension, hydraulic brakes and a four-speed synchromesh gearbox. Its rear-mounted engine, an air-cooled vertical twin 393 cc unit, could power the car to a top speed of 50 mph (80 km/h). The bodywork was unitary.

It will be readily ascertainable that the Vespa was a well-finished and competent little car streets ahead of most other microcars. Considering it was competing with Fiat's 500, it is a great credit to the Vespa that, up until 1961, about 34,000 cars were built.

Although instigated by the famous Italian motorbike firm Piaggio, the pretty Vespa 400 was actually built in France.

ITALY

SIATA

Siata of Turin is one of the great names in Italian automotive history, first as engine tuners and latterly as purveyors of some handsome coachbuilt coupés and tourers. Siata's microcar exploits are less well-known.

The first production car from Siata was the Amica, a Fiat 500B/C based cabriolet, introduced in 1948. In 1953 came the firm's attempt at a 'mass-market' car, the Mitzi. This was a very basic two-to-three-seater coupé initially fitted with a 398 cc two-stroke twin in the back. Production versions were fitted with a four-stroke engine of 434 cc capacity. Suspension was by torsion bars all round.

Turin, April 1954: the 400 cc Siata Mitzi makes its début alongside Siata's rather more exotic offerings.

Weighing 890 lbs (405 kg), the Mitzi was capable of a top speed of 52 mph (85 km/h). Siata turned out to be more interested in, and capable with, more expensive coachbuilt motor cars and production of the Mitzi was abandoned after just a couple of years. Only around 50 had been made, although more were made under licence in Argentina under the name Ryca.

In later years, Siata made more of a go with the pretty Fiat 600-based Amica Coupé and Cabriolet. There was even a version fitted with a Fiat 500 engine called the Nuova 500 Spider.

PANTHER

One of the most sophisticated and intriguing microcars of the 1950s was the Italian-built Panther. It was unique in having a twin-cylinder diesel engine of 520 cc and plastic bodywork — no other Italian car used plastic in its bodywork at that time — and front-wheel drive, again unusual for an Italian car.

The first model of 1954 was called the Colli and was a fairly attractive looking coupé claimed to offer 50 mph (80 km/h) and 90 mpg. A petrol-engined version was also listed (using an 18 bhp 480 cc four-stroke unit). A rather prettier Zagato-bodied Panther was another variation in 1955.

Licensed production was due to happen in Belgium, France and Argentina and a company was formed to manufacture the Panther in the tiny principality of San Marino, but nought came of any of these projects. However, San Marino should not be struck off the car manufacturing map. It later became the home of another microcar: the DECSA Lisa of the 1980s (see Chapter 5) — and also, incidentally, an unusual retro-cabriolet called the Epocar Symbol from 1990.

The Italian-made Panther is a rare example of a diesel-engined microcar.

NETHERLANDS

SHELTER

Microcars from the Netherlands are a rare breed, especially in the 1950s, but perhaps the most promising was the Shelter. Designed by Mr van der Groot, the first Shelter was made in 1947 and a number of prototypes, all with three wheels, were built up to 1955, by which time a company had been formed with the intention of producing a car.

This was a very small closed three-wheeler which was to use a specially built 228 cc engine (prototypes had a 200 cc Ilo engine), operated via a three-speed gearbox. Perhaps the most remarkable fact about the Shelter was that virtually the whole car, including the drivetrain, was designed by van der Groot.

In 1956, the Dutch Government gave financial assistance to the project and four cars of a planned batch of 20 were built, which would be hired to customers. But before the project could develop, the Government finance was withdrawn (it was now investing in DAF) and the Shelter was left out in the cold.

One example of only a handful of Shelters made by Dutch inventor van der Groot.

This Austrian Felber Autoroller travelled hundreds of miles to attend the 1991 UK National Microcar Rally.

AUSTRIA

FELBER

The Austrian car industry is hardly world-renowned, but it did produce a few microcar firms during the 1950s. Ernst Marold's Felber marque was, for a while, successful in making the Autoroller. This was a small convertible three-wheeler with a canopy which lifted up for entry. Its two-cylinder Rotax engine sat beside the single rear wheel. Initially, the unit was 350 cc and developed 12 bhp, but this soon increased to 398 cc and a rorty 15 bhp. The prototype was completed in 1952 and the Vienna-based firm supplied around 400 cars before production ceased in 1954. Only two Felber Autorollers are known to exist today.

CHAPTER THREE

THE SURVIVORS

THE turn of the decade into the 1960s saw the turning of the tide for the bubble car. In the watershed which followed, the myriad bubbles of Europe burst into nothingness. One by one, the great names of a heady boom fluttered out so that, by 1965, there was hardly a familiar face left.

The reason was the unstoppable advance which mass production small cars had made. The Mini was the bubble-burster par excellence in Britain, while in Italy, the Fiat 500 came to rule supreme in the congested cities. All across Europe, better small family cars manufactured by NSU, Simca, Citroën, Renault and others made travelling by bubble look like trouble. An increasing affluence made the austerity of the majority of microcars unpalatable to most.

But there were survivors, albeit a pitiful few. As most were swept away by the tides of change, those able to occupy a safe, small niche on the cliff face stood a good chance of survival. In Britain, there was actually a steadily increasing demand for three-wheelers over the decade of the 1960s thanks to the tax and licensing laws, reaching a peak of some 13,000 units in 1972.

Those benefiting from this trend included Reliant and Bond, by far the largest British manufacturers of tripeds. Bond was eventually taken over by Reliant, leaving the latter the sole defender of the triking market in the UK into the 1970s.

Abroad, there were other niches. In Greece, for example, the huge taxes imposed on imported cars gave the Greek version of the Fuldamobil an extended lease of life — until 1977, in fact. And in Czechoslovakia, the cheap transport offered by the Velorex kept it churning out until 1971.

But it was a new breed of car which took the microcar on to new ground. With the cultural explosion of the 1960s, there was a certain demand for transports of delight — pure fun cars. A market opened up for follies in miniature. In Italy, this took the form of innumerable jeeps, buggies and cute convertibles based on the Fiat 500. In Britain, there was a fun car boom from 1968 to 1972, during which time many small fun machines were launched in kit form, usually based on the Mini.

There were also the seeds of a future generation of micros in Italy, where tiny cars like the Zagato Zele and Lawil found an avid market. During the late 1970s, this type of car was to blossom with incredible diversity, particularly in France, in a new microcar boom: for this, see Chapter 4.

GREAT BRITAIN

RELIANT

Reliant was Britain's biggest supplier of three-wheelers as the bubble car boom came to its close. It managed to escape the collapse by having forged strong links with the ex-motorcycle driving family man, who has remained the saviour of the Reliant three-wheeler right up to the present day.

Reliant had taken over production of the Raleigh three-wheeled van from 1935 but its first passenger three-wheeler was the Regal of 1952. It used Reliant's own 16 bhp 747 cc version of the Austin Seven engine in an open four-seater metal body. The body material switched to glassfibre in 1956 (after which Reliant never looked back — all its cars were henceforth made with plastic bodies). The body style changed at this point, too, to a more rounded shape and a two-door saloon version was offered alongside the tourer.

The Regal progressed through six Mark designations, adding such refinements as a curved windscreen, synchromesh for the three-speed gearbox and 12-volt electrics. The last Regal Mk VI rolled out of Reliant's Tamworth factory in 1962.

Reliant established itself as Britain's leading manufacturer of three-wheelers with the Regal.

Its replacement was the Regal 3/25 (3 wheels, 25 bhp). It had the distinction of being fitted with Britain's first mass-produced aluminium engine, developed by Reliant: a 598 cc die-cast alloy four-cylinder unit with a creditable 25 bhp on tap. Its glassfibre body also aped current trends by incorporating a Ford Anglia style 'breezeway' rear window; the effect was, however, hardly handsome. An estate version was also offered.

In 1964, the Regal was joined by the four-wheeled Rebel, a more conventional looking car which shared the Regal's mechanical basis; again, an estate version was offered. Both models received Reliant's expanded 700 cc engine in 1967 and got a modest performance boost from the extra power (31 bhp). The Regal then became known as the 3/30. Reliant's 50,000th Regal was built in April 1968.

Both the Regal and Rebel continued into the 1970s, the Rebel with a 748 cc engine for the last year or so of its life: it was withdrawn in

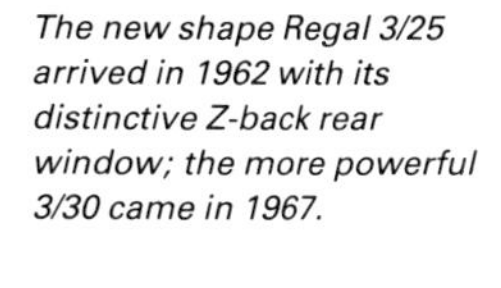
The new shape Regal 3/25 arrived in 1962 with its distinctive Z-back rear window; the more powerful 3/30 came in 1967.

This Reliant Regal GT(!) was offered by Twin Strokes and featured an extra 1 bhp, a matt black bonnet and streamlined wing mirrors...

December 1973. The Regal was replaced in 1973 by the Robin, a much more modern-looking GRP three-wheeled saloon with bodywork styled by Ogle and a new chassis. It used the 748 cc engine (72 mph) until 1975, when a new 848 cc unit was standardised; power was now up to 40 bhp. Estate and van versions were also offered.

The Robin was by this time the last mass production three-wheeler sold in Britain and came in for a certain amount of stick because of it. Police called them 'plastic pigs', but the Robin continued to attract buyers oblivious to the jokes. It was joined by a four-wheeled version, called the Kitten, in 1975, which impressed with its excellent economy: up to 50 mpg. The Kitten, too, was available with estate bodywork.

In a chapter of survivors, Reliant out-survived them all: the Kitten was deleted in 1982 (having sold 4,074 examples), and the Robin's replacement, the Rialto, is still available at the time of writing, albeit under the auspices of Beans, which bought up all the tooling following Reliant's bankruptcy in 1990.

Reliant's Rebel of 1964 was a four-wheeler with an all-new glassfibre body; 70 mph and 60 mpg were claimed.

Ogle-designed Robin replaced the Regal in 1973, offering more power from its enlarged 748 cc engine.

BOND

One of the reasons why Bond was able to survive the great microcar crash was its expansion into specialist sports car production with the Equipe from 1963. In this respect it mirrored its future owner, Reliant. But Bond did not abandon microcar production.

The final 'traditional' Bond had been made in 1966 (see Chapter 2). From 1965, a new model became available in a rather different vein: the new 875 three-wheeler used a low compression (de-tuned) Hillman Imp engine of 875 cc capacity in the rear. Although developing only 34 bhp, this gave it a top speed of 82 mph and 0-60 mph in 16 seconds: definitely more sporting than any previous Bond three-wheeler.

If there was one thing the Bond 875 was not, *it was beautiful. Bond's ad campaign brought back memories of its slogan for the Equipe: "Is this the world's most beautiful car?"*

The 875 had an enclosed four-seater glassfibre body which was also rather more accommodating than traditional Bonds. Ranger Van and estate versions were also available. A Mk 2 version of the saloon in 1967 had rectangular headlamps, but the model died when Reliant closed down the old Bond factory in Preston in 1970, having bought the company the previous year.

That was the last of the 'true' Bonds. Reliant used the Bond name on a new fun car which it launched in 1970. Instigated by Tom Karen of Ogle Design, with whom Reliant had a close liaison, the Bond Bug was a bold design combining the economies of three-wheeled motoring with a sporting, youthful image.

A pre-production prototype of the Bond Bug: note the Porsche 928-style flop-back headlamps and rear indicators on stalks.

The Bug (seen here in 700E guise) was aimed squarely at young drivers "who know how to enjoy life."

Its most striking feature was a one-piece canopy which tilted forward to allow entry; removable sidescreens provided ventilation. There were two seats and a small amount of luggage space in the sharply cut-off tail. The wedge shaped body was available in any colour you liked — as long as it was bright tangerine! The use of aircraft-style black decals was also revolutionary.

Initially it was intended to be sold in three versions: the 700, 700E and 700ES, although in the event the 700 never entered production. The ES used the high compression (31 bhp) version of the Regal/Rebel engine, sufficient to power it to a top speed of 77 mph. Lesser Bugs used the 29 bhp engine. From October 1973, Bugs received the bigger 748 cc engine and the models became known as the 750E and 750ES. Production came to a close in May 1974 with around 2,270 sold. The very last cars were registered in 1975.

MINK

A more saucer-like microcar than the Mink has never been built. It was made by a Midlands engineering firm in 1968 for a company which hoped to find a market for a production run in Bermuda.

Using a 198 cc Lambretta scooter engine sited in the rear, the Mink was capable of 55 mph and 70 mpg. Weighing only 530 lbs, its motor scooter wheels and shock absorbers contributed to a tendency to oversteer violently. Built on a backbone chassis, the tiny glassfibre body featured a precipitous front overhang in the effort to retain a saucer-like shape. There was no reverse gear, but the Mink's rear end was very easy to manoeuvre into a parking space. It is believed that the Mink remained a prototype.

The TiCi 'city sprint commuter car' used a rear-mounted Mini engine and was exceptionally short.

TICI

Anthony Hill was a lecturer in furniture design who decided he should make some mobile sculpture. His first attempt was a small enclosed two-seater using a 500 cc Triumph Daytona motorbike engine mounted in the rear. It took three years to build and, when it was finished in 1969, he called it the TiCi (pronounced 'tichy'), which it certainly was: a mere 6 feet long and weighing only 560 lbs.

But it wasn't the right sort of car for the age. The arrival of the fun car boom in Britain gave Hill a chance to market a larger version with a Mini engine from 1972. Described as a 'city sprint commuter car', it was still only 7ft 5ins long but weighed rather more at around 1,000 lbs.

Sold in kit form for £395, the open doorless body could be ordered with a hard top and doors as extras. Looking very squat and angular, it had a certain chic appeal: the singer Eartha Kitt and the pop group Showaddywaddy both bought one and Stirling Moss promoted the car. The author also owned the prototype.

This wasn't enough to save it from extinction, along with most other fun cars in 1973. 40 examples had been made, of which 12 were exported to Spain and Japan.

ITALY

FIAT

Fiat's domination of the Italian small car market after the war was down to its brilliant Topolino, which it had been making since before the war (see Chapter 1). It was this car which prevented very much of an Italian microcar industry from happening. The model which took over its mantle was the Fiat 600, which was even smaller than the Topolino.

The 600 was not a great car but it did offer a full four seats and reasonable performance from its rear-mounted 633 cc engine — something most other small cars could not offer. From 1955 to 1960, the 600 sold a remarkable 891,107 examples. There was also a 600-based forward control people-carrier called the Multipla (built from 1955-66). This six-seater truly initiated the 'space van' theme. The 600 saloon was replaced in 1960 by the 767 cc-engined 600D, which is getting a bit too big for a book about microcars. However, the true replacement for the Topolino, the Nuova 500, makes a different story.

What Fiat created when it designed the 500 is often dismissed as a crude and sluggardly oddity. In reality, it was inspired. Its hidden charms have been realised by such owners as Lotus and McLaren stylist Peter Stevens, who drives one to work every day.

The 'new' 500 arrived in 1957. It had modern-looking bodywork sporting 'suicide' (rear-hinging) doors which fitted four people (just). 500s also had a welcome folding sunroof.

In its original form, it had a 479 cc air-cooled vertical twin-cylinder engine mounted in the rear. This developed a rather feeble 13 bhp, sufficient for a top speed of 53 mph. The four-speed gearbox was entirely non-synchronised. There was independent springing on all four corners, though, and a 'Cinquecento' was great fun if you weren't in a hurry. From 1958, there was even a 500 Sport with a bored-out 499.5 cc engine developing 20 bhp.

Fiat's 600 was a practical successor to the 500 Topolino with four full seats and a rear-mounted 633 cc engine.

Based on the 600, Fiat's Multipla was a genuine six-seater – remarkable for a car only 139 inches long.

The Nuova 500 of 1957 was a revelation: the most successful of all microcars, selling some 3 million examples.

The 500 Giardiniera estate cleverly used the 500's rear-mounted engine tilted sideways to clear luggage space.

In 1960, the larger engine became standard with the 500D, albeit with 17.5 bhp on tap. Its top speed was now 58 mph. Gone were the suicide doors, too. In the same year, Fiat introduced an estate version called the Giardiniera, which cleverly allowed for generous luggage space behind the rear seats by mounting the new engine on its side underneath the floor.

Although it was a slow seller at first, in time the 500 became Italy's archetypal city car: at times in the Milan rush hour, you could be surrounded by a sea of 500s and very little else. The 500 also spawned a huge industry in alternative bodies using the mechanicals, or the floorpan, of the small Fiat. From famous names like Autobianchi and Vignale to the highly obscure (like the off-road Ferves), it was paradise for microcar lovers.

There was a huge industry in go-faster versions of the 500 from companies like Abarth (up to 38 bhp and 90 mph). Tuners Giannini even offered an Economy Run version with a 390 cc engine for better fuel consumption (70 mpg) — as if that was really necessary. Ghia built a fabulous-but-frivolous open beach car version called the Jolly from 1957 (other Jollies were also built on the 600, Multipla and Giardiniera).

In 1965, the 500F was announced with slight improvements to specification; once again, in 1968, the 500L added more refinements. After Fiat took control of Autobianchi in 1968, production of the Giardiniera estate was transferred over to that division and the car was henceforth badged as an Autobianchi until its demise as late as 1977.

The 500 saloon lived on two years beyond the introduction of its intended replacement, the 126, in 1973. Some wished that Fiat hadn't bothered to replace it, as the boxy new 126 had none of the charms of its jellymould predecessor. When the 500 finally bowed out in 1975, nearly 3 million had been made.

The 126 looked very similar to the 500 on paper: it had a rear-mounted twin-cylinder engine, four seats and a very cheap base price. In reality, it was a watered-down concept: it was considerably larger than the 500 (at over 10 feet, or 305 cm, long), without offering significantly improved interior space, did not have a sunroof as standard and was, in short, not as much fun. Its 594 cc engine (23 bhp) did give it slightly better performance: Fiat claimed 65 mph (105 km/h) for the little car.

As the years progressed, the 126 got a larger 652 cc engine in 1977 and a plush De Ville version was introduced the same year. Production was transferred in 1987 entirely to Poland (where licensed production of the 126 was already well established). In 1987, the revised 126 Bis, with a 704 cc water-cooled twin replaced the old air-cooled unit and wheel size was increased. A hatchback also became a new fitment. The 126 was for many years the cheapest car available in the UK, but the last examples were sold in the summer of 1992.

Production continues, however, in Poland, where a revised version was introduced in late 1994. To date, well over 2 million examples have been built.

AUTOBIANCHI

Autobianchi began making cars after the war when Fiat helped finance a project based on its new Fiat 500. The Autobianchi Bianchina arrived the same year as the 500, 1957, and was a 'rolltop' coupé using the 500's mechanical parts. But it was a paragon of luxury compared to the Cinquecento: it had lots of chrome and a better trimmed interior which retained seating for four.

From 1958, it got the option of the new 20 bhp 499.5 cc engine, which gave it a top speed of 62 mph. In 1960 came the Bianchina Special Cabriolet, a full convertible with minor styling revisions. And to coincide with Fiat Giardiniera, Autobianchi introduced its own estate based on that model: the Panoramica. This was also available with blanked-off side windows as the Furgoncino ('little truck'). The final bodywork option of the Bianchina arrived in 1962, when the Qattroposti ('four-seater') saloon was launched. In some countries, the Bianchina was marketed with the inappropriately cosmopolitan name of Eden Roc!

In 1963, the Coupé was deleted but the other models continued on until 1969; the van made it until 1970. Autobianchi was completely taken over by Fiat in 1968 which decided that Autobianchi should make the Giardiniera, which it did until the model's death in 1977.

There was only one further micro Autobianchi: the Stellina sports convertible of 1964. This was a two-seater based on the Fiat 600, but it was hardly a pretty car and made virtually no impact.

Based on the Fiat 500, Autobianchi's Bianchina was available as a saloon, coupé, estate, van and convertible.

The Vignale-Fiat Gamine was a frivlous little retro-roadster based on a 500 floorpan; inevitably it was nicknamed the Noddy car.

VIGNALE

Alfredo Vignale's Grugliasco-based coachbuilding works was one of the early builders of special bodies for Italian cars, notably Ferrari, Maserati and Fiat. Among its many small car designs were new coupé and cabriolet bodies for the Fiat 600 and 850.

But its smallest ever car was the Gamine, a quite ridiculous machine based on a Fiat 500 floorpan. Its styling was a supposed pastiche of the pre-war Fiat Ballila but in Britain it was inevitably called the 'Noddy car', a tendency not helped by the bright tangerines and peppermint greens the cars were painted in.

The Gamine arrived in 1967, aimed squarely at Mediterranean 'Belle Epoque' swingers, with whom it scored a certain hit. All its styling details were designed for amusement, from the froggy headlamps and 'traditional' fake grille to the cut-away doors and rounded rump. And of course it was a convertible.

It shared the Fiat 500's technical specification and was equally short (only 9ft 10ins). But it was merely a two-seater and was reportedly not particularly well-built. Its days in Toy Town were numbered: Alfredo Vignale sold out to DeTomaso in 1969 and the Grugliasco factory was soon pressed into service constructing the powerful DeTomaso Pantera sports GT.

One intriguing side avenue to the story was the importation of several hundred Gamines into the UK by a Greek Cypriot entrepreneur named Frixos Dimitriou, from 1968. A casino owner, he attempted to set up a more respectable enterprise by buying up huge quantities of cars from Vignale — even though he had no buyers for them. As a consequence, most of the Gamines brought over to the UK were re-exported to Cyprus. Mr Dimitriou ended his days in another Vignale when a runaway tank in Cyprus squashed both him and the car flat.

GREECE

ATTICA/ALTA

The German Fuldamobil (see Chapter 2), like so many successful micro-car designs, was made under licence in numerous countries. But in Greece, it managed to stay in production for an inordinate number of years.

Perhaps the most impressive survivor of the bubble boom was the Alta from Greece: this 1950s design lasted until 1977.

The Greek plastics manufacturer, Georgios Dimitriadis, took on licensed production in 1964, where the car was known as the Attica. The three-wheeler used a 198 cc Heinkel single-cylinder engine (top speed 60 mph). In 1966, it was joined by the open Cabrioletta version. About 100 saloons and a handful of convertibles were made before the manufacturing firm had to sell the project on to a new firm, Alta, in 1968.

Alta modified the design, including more angular rear bodywork for 2+2 seating and fitted a 198 cc Sachs engine (10 bhp). In this form, it became one of the stalwarts of the Greek motor industry but, even in that protectionist market, the Alta's attractions steadily faded. Even so, the last Altas were made as late as 1977. Alta also made an open three-wheeler and a larger BMW 700-engined enclosed three-wheeler. Mr Dimitriadis went on to build the DIM in 1977 (see Chapter 5).

AUSTRIA

MEISTER

Austrian car manufacture is an obscure subject. The Meister G5N Mopedkabine is an obscurity within the obscure. Produced by Hans Meister of Graz in 1969, it was a tandem two-seater three-wheeler with the most basic plastic bodywork over a tubular frame. Steering was by handlebars. The engine was a 49 cc Puch 3.5 bhp moped engine, only sufficient for a top speed of 25 mph (40 km/h). It drove only one rear wheel and the driver had to kick start it into life.

The Meister G5N from Austria was nothing more than a glorified moped, but at least you got protection from the elements.

The infamous Trabi 601 belched its two-stroke way into the record books, lasting from 1964 to 1991 and selling in its millions.

EAST GERMANY

TRABANT

Everyone knows what a 'Trabbie' is. In the western world, it has achieved a kind of cult status following the collapse of the Berlin Wall. In its own country, the ex-East Germany, it is largely a shunned object, as the enforced driving of Trabants has been eased by the ready availability of second-hand BMWs and Mercedes.

The first Trabant grew out of the Zwickau works, itself derived from

The East German Trabant first appeared in 1958 with rounded pressed plastic bodywork.

the fragmented Audi and DKW firms. It appeared in 1958 as the Trabant P50 and effectively became the 'Volkswagen' of East Germany.

Its specification was crude: a 500 cc 18 bhp two-stroke twin encased in a body made of duroplast, a textile reinforced plastic that was pressed, not moulded, into shape (although there was a steel floor). There were saloon and Kombi versions. The engine grew to 600 cc in 1963, but the following year, the model was replaced with the more squarish 601.

The remarkable thing about the 601 is that it was a survivor *par excellence*, lasting right up until 1991. It kept the same duroplast construction – earning it the nickname 'the cardboard car' — and soldiered on with the same 594 cc two-stroke engine until May 1990, when it was replaced by a 1.1-litre VW Polo unit. The environmental impact of the two-stroke engine is very visible in the blackened buildings in the former East Germany.

The saloon and Kombi estate version were also joined in 1967 by a civilian version of the military convertible Trabant, called the Tramp (which today has some chic value). There were even tuned and bored-out versions offering up to 70 bhp for the mightily brave.

Throughout its production life, the Trabant had a 12-15 year waiting list. Today it is universally unloved. A remarkable total of around 3 million Trabants were made from 1958 to 1991.

CZECHOSLOVAKIA

VELOREX

If the Trabant was an unusual 'people's car', its Czechoslovakian equivalent was yet more outlandish. The Velorex was another great survivor, lasting from 1954 until 1971.

Perhaps the strangest facet of the Velorex was its canoe-like construc-

The Czechoslovakian Velorex was bodied in canvas over a steel frame; amazingly, its production life extended into the 1970s.

tion. Over a steel tube spaceframe, canvas panels were stretched and then fastened. The front and rear panels had stud fixings so that owners could remove them. A de-luxe version with all-steel panels was offered for those with misgivings about canvas bodywork.

Velorex also made motorbikes and took its engines from the Jawa range: 250 cc single-cylinder and 350 cc two-cylinder two-strokes were used to drive the single rear wheel. A four-wheeled model also made an appearance but did not enter production.

AUSTRALIA

LIGHTBURN ZETA

In such a vast country as Australia, it is perhaps surprising that anyone should attempt to make a microcar, but several have done so: from the Edith and Hartnett to the Tilbrooks and Buckle Dart, there have been a brave few.

None was so ambitious as the Lightburn Zeta. Its story begins in Britain in 1958, when the makers of the Frisky, Henry Meadows, displayed the sporting Friskysprint. In contrast to the small microcar they were already making, the Friskysprint was low, sleek and beautiful, styled by Gordon Bedson and had a 492 cc Excelsior engine mounted in the tail.

However, Meadows did not build the Friskysprint in series (see Chapter 2) but sought a buyer for the project. One was found in Lightburn of Adelaide, Australia, makers of boats, concrete mixers and car jacks. Instead of the unreliable Excelsior engine, Lightburn knew that FMR, the makers of the Messerschmitt, had some surplus stock of 493 cc Tiger engines and obtained a batch to fit into their new car.

In this form, the new car — called the Zeta Sports — weighed only 8 cwt (407 kg) and was claimed to do 75 mph (120 km/h) and reach 0-50 mph in 12 seconds, which was very sportscar-like for the era.

Lightburn displayed the Zeta Sports as early 1961 but production did not begin until 1964. By this time, they were also launching a micro utility car simply called the Zeta.

The Zeta saloon was the opposite of the Sports: very tall and dumpy with a two-door estate-type body, again in glassfibre. It was practical, however: the four seats could be folded or removed for extra luggage space or even mounted on to the roof for a grandstand view at sports events! But there was no rear hatch or door...

The engine was, perhaps not co-incidentally, the same as the Frisky-sport's: the 16 bhp 324 cc Villiers two-stroke, as made in Australia. This drove the front wheels via a four-speed gearbox. The engine was also reversible, allowing the Zeta to travel at 60 mph *backwards* as well as forwards. It acquitted itself well in some endurance competition events; at one point it even floated while crossing a river!

There was not much demand for either of Lightburn's models. Plans to build 50 cars a week quickly foundered and, by the time the firm ceased making cars in 1967, some 363 saloons and pick-ups and just 49 Sports models had been made.

The Zeta Sports from Australia was powered by a 500 cc Messerschmitt Tiger engine, promising "back-thumping acceleration."

CHAPTER FOUR

JAPAN'S 'K' CLASS

WHEN your traffic jams are in danger of becoming full-scale preserves, you are forced to do something drastic. The traffic traffic jams of Tokyo and Osaka are the worst in the world; and the Japanese government's response is one of the most pragmatic in the world.

A government agency sets strict size and specification limits for distinct classes of vehicle. The *Kei-jidosha* class (the 'K' class) is the smallest group. Initially, these guidelines called for a class of vehicle which was no longer than 3.0 metres (9 feet 10 inches) and with an engine capacity of no more than 360 cc. Progressively, the limits were upped to 3.2 metres and 550 cc and, today, to a maximum of 660 cc, with a power output of 63 bhp, a maximum length of 3.3 metres and a maximum width of 1.4 metres (55 inches).

Tax allowances on this type of microcar were just one benefit of owning a 'K' car in Japan. More significantly, you were allowed to park your car on city streets at night; owners of all other types of car had to prove they had a parking space before being allowed to own one.

From the very beginning of the class in 1955, these 'K' cars showed great ingenuity in cramming as much car as possible into a very small space. Most had four seats and often a hatchback. Some were offered in estate form, or as pick-ups, even jeeps. Others manufacturers were brave enough to attempt coupés like the Mazda R-360 and even sportscars, like the Honda Beat.

As the category became established, an increasing sophistication became apparent. The Honda N360 got an automatic transmission option in 1967; and Honda's Z600 Coupé of 1970 was, in style and specification, decidedly grown-up, incorporating front disc brakes and a five-speed gearbox in its most up-market guise.

The class became increasingly popular, reaching a peak in 1970, when 750,000 were sold. Through the 1970s, the tax and other benefits of the class were eroded by government legislation and sales suffered to the extent that manufacturers insisted on relaxed government constraints to make the miniature class more saleable. By 1980, this seemed to have worked and sales once again exceeded 700,000. In 1993 some 772,368 'K' class cars were sold in Japan, and the combined microcar/van class accounted for 24% of the entire Japanese market. Suzuki is the market leader, followed by Daihatsu, Mitsubishi and then Honda.

In the 1990s, the 'K' car is an established and integral part of motoring in Japan. 'K' cars are uniquely able to cope with city traffic and offer unparallelled economy motoring. The current crop of these cars is highly sophisticated, incorporating turbochargers, superchargers, four-wheel-drive, four-wheel-steering, air bags, air conditioning, electric windows and much more. Some are real performance machines, able to accelerate from 0-60 mph in around 8 seconds.

The only Japanese firms never to have been involved in the 'K' class are the giants Nissan and Toyota, plus Isuzu. Every other manufacturer currently has a contender in the market place. However, most reason that the genre is too specialised and too unprofitable for export markets. In Britain, only Honda and Daihatsu have been brave enough to try importing genuine 'K' class cars. After Honda's disastrous experience with the poor-selling Jazz (only 400 were sold in the UK), it's unlikely that it will ever bring in its delightful Today. Recently, Subaru and Suzuki have joined Daihatsu in importing 'K' cars to Britain with the Vivio and Cappuccino respectively.

The continued existence of a special microcar class in Japan has largely prevented the growth of any other type of microcar in Japan. In the 1950s there were a few brief sub-'K' attempts but, since then, the numbers can be counted on the fingers of one hand.

The future of the *Kei-jidosha* class in Japan is in doubt. In 1994, the Government announced its intention to strip the microcar class of its last remaining privilege – the tax advantages offered to it – so that the 'K' class cars would have to compete on their own merits. This decision was reported to be a result of the class's commercial success: there were simply too many microcars being sold!

On the other hand, the existing Japanese microcar manufacturers are lobbying the Government to increase the limits of the microcar class to 800cc and larger dimensions. Whether the 'K' class can regain its status remains to be seen, but the question must be asked: can gridlocked Japan really afford to be *without* them?

MAZDA

Mazda launched its car production existence in 1960 with a microcar: an attractive and diminutive coupé called the R-360. It was powered by a vee-twin air-cooled engine of 356 cc capacity. Producing just 16 bhp, it propelled the tiny (870 lb) 2+2-seater to a top speed of 56 mph. Remarkably, it proved an instant sales success, scoring 23,417 sales in its first year.

Mazda's pretty R-360 coupé was an immediate hit from its launch in 1960.

This allowed the quick introduction of the B360 pick-up in 1961 and the Carol 360 in 1962. The latter used a water-cooled rear-mounted four-cylinder engine of 358 cc capacity, producing 20 bhp and powering the little car to a top speed of 59 mph. Both two and four-door versions were offered. The model shared the success of the R-360, capturing an incredible 67% of the micro class in its first year of production and catapulting Mazda into the motorcar big-time. In 1964 came the non-'K' class 586 cc Carol 600.

The Carol was not replaced until 1972, when the Chantez was launched. This reverted to a two cylinder engine, although it was an in-line water-cooled unit of 359 cc displacement, placed up front. Its 35 bhp output was enough to give it a top speed of 71 mph and a standing quarter mile in a surprisingly nippy 20.6 seconds. There was only one body style, a two-door saloon, which betrayed the fact that Mazda's heart was no longer in microcars, preferring to develop its expanding range of larger models. The Chantez died a natural death with no replacement.

The Carol name was revived for Mazda's re-entry into the 'K' class arena in 1989, when Mazda felt sufficently inspired by the revival of the 'K' class in general. The new Carol (actually sold under Mazda's

The Carol was the car on which Mazda founded its fortunes: for 1961, it was easily the best 'K' class car on sale.

Autozam marque) was a whacky, boldly curved design notably using circular headlamps and semi-circular window glass. It borrowed its 547 cc three-cylinder engine and the rest of its mechanicals from the Suzuki Alto. Soon, a variety of options became available, including a full-length sunroof, four-wheel-drive and turbocharging. Alongside other 'K' class cars, the Carol's engine was uprated to 657 cc (from Suzuki) in 1990. The most powerful Turbo version now developed 63 bhp.

In 1992, it launched another brave 'K' car, the AZ-1 sports car based on the AZ550 Type A prototype shown at the 1989 Tokyo Motor Show. This was a superb two-seater coupé using the most powerful (63 bhp) version of Suzuki's three-cylinder 657 cc engine mounted amidships. Entry was via heavily glassed gullwing doors — the first 'K' car to have these. Like almost all 'K' cars, the AZ-1 remained a car sold only in Japan. A clone version was also made for Suzuki under the name Cara, with different badging and a restyled nose. The exchange of microcar technology was extended in 1994 when Mazda asked Suzuki to supply it with badge-engineered versions of the successful Wagon R.

1989 saw Mazda's re-entry into the microcar arena after a gap of over ten years with the eccentric little Carol.

Amazing, individual, impractical – the tiny gullwing Mazda AZ-1 was a mid-engined 'K' car offering every bit as much fun as a Honda Beat.

SUZUKI

Suzuki was the first firm to launch a 'K' class car with the front-wheel drive Suzulight of 1955. It had an air-cooled two-stroke twin of 360 cc capacity but a mere 43 examples were made. The new generation Suzulight 360 of 1959 was prettier, more modern and far more successful, establishing Suzuki's name as a car manufacturer, not just a motorbike maker. Top speed was 53 mph (85 km/h). An estate version was also available.

The Suzuki Fronte 360 of 1967 was typical of the era: a rear-engined two-stroke engine and a top speed of up to 78 mph.

Its more modern replacement in 1967 was the Fronte 360, the first of a long line of microcars bearing that name. Its air-cooled two-stroke three-cylinder engine was mounted in the tail and was available in two states of tune: 25 bhp and 36 bhp. The latter version, known as the SS, could reach 78 mph. Alongside it, a novel estate version, known as the 360L 20V, offered people carrying capacity in miniature.

Suzuki's curious Fronte saloon of 1975: an odd amalgam of styling influences. This is the mid-range FC 4-door.

The even weirder Fronte Hatch was a tall 2+2 two-door intended as a luggage carrier.

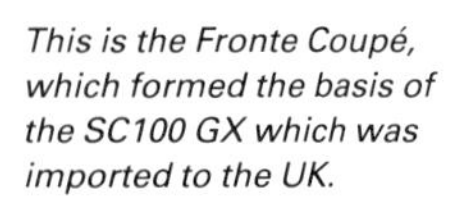

This is the Fronte Coupé, which formed the basis of the SC100 GX which was imported to the UK.

It was replaced in 1970 by the new Fronte, which shared the outgoing model's engine and mechanical specification, but had a rather plain all-new body. In later versions, it became typical of the attempt to impose American styling influences on the constrained medium of the sub-3 metre-long car: it looked rather strange in all its guises, and there were several: two and four-door saloons, a coupé and an odd three-door estate car called the Fronte Hatch. But it was a good example of how sophisticated the breed was becoming: its standard equipment included a radio, clock, cigar lighter and rev counter.

It is a little known fact that the ubiquitous Suzuki 'jeep' began life, and continues today, as a 'K' car based on the Fronte. In Japan it has always been known as the Jimny, right from the moment it was first introduced in 1970. Then it had a 28 bhp 359 cc two-stroke engine, sufficient to get it to just 48 mph. It was a unique 'K' car in one respect: it had four-wheel-drive and was a genuine off-roader. As the years rolled by, the engine options expanded to 539 cc, then 797 cc, then 970 cc and even 1298 cc. All the while, there has remained a basic version which conforms to Japan's *Kei-jidosha* size and engine restrictions, even today.

The Cervo 2+2 coupé of 1972 was again based on the rear-engined Fronte but used a 539 cc two-stroke engine developing 28 bhp, in which form it could reach just 65 mph. It was a far cry from the rumbustuous export model, the SC100 GX, which was known in Britain by its nickname, Whizzkid. This had a 970 cc four-cylinder engine developing 47 bhp, enough for a top speed of 85 mph and sparkling acceleration. This was the same engine fitted to the export model of the new front-wheel drive Fronte of 1979, which was known as the Alto in Britain.

In 1982, Suzuki redesigned the Cervo around the front-wheel-drive Fronte/Alto, using its 30 bhp three-cylinder 543 cc engine. Racier versions followed, including the DeTomaso Turbo (1983) with 40 bhp on tap and, in 1984, the CT-G Turbo with 48 bhp and 81 mph. Also in 1983 came a cute two-seater pick-up version of the Cervo delightfully called the Mighty Boy which was, for a time, the cheapest car available in Japan and, perhaps, anywhere in the world (at an equivalent of around £1200).

Its successor was the Cervo Coupé, based on the new (1979) Fronte. This is the CT-G Turbo model.

The cheapest mass-produced car sold in the world in 1983: the Suzuki Mighty Boy two-seater pick-up.

The increasing refinement of 'K' cars was reflected in the Fronte's changing body styles (in 1985 and 1988) and the development of four-wheel-drive and a fuel-injected twin-cam 'Works' turbo engine, providing 63 bhp. As the Fronte name was dropped in favour of the Alto tag, engine size was upped to 647 cc. The Works 660RSR continued the performance trend. For export markets, the Alto was usually fitted with a three-cylinder 795 cc engine.

In response to Honda's Beat, Suzuki launched its own 'K' class sportscar in 1992. Called the Cappuccino, it was a brilliant front-engined, front-wheel-drive, open two-seater with a three-cylinder 657 cc engine.

Suzuki's most successful K-car came in 1993 with the amazing Wagon R. Attempting to get the most space out of the dimensions allowed, the roof of the Wagon R was high enough to accommodate four top-hat wearers. It was characterful in a gawky sort of way and really caught on with Japanese buyers, spurred on by its winning the Japanese Car of the Year Award. It was made at a rattling rate, entering the Top Ten table for the best-selling cars in Japan.

Suzuki has therefore remained probably the most ardent advocator of the 'K' class car in Japan, with a model range spanning an unbroken 38 years to date.

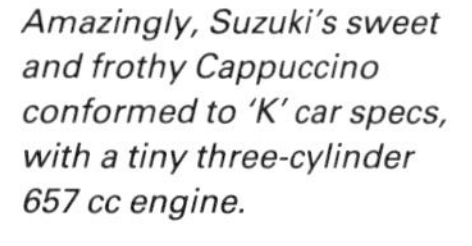

Amazingly, Suzuki's sweet and frothy Cappuccino conformed to 'K' car specs, with a tiny three-cylinder 657 cc engine.

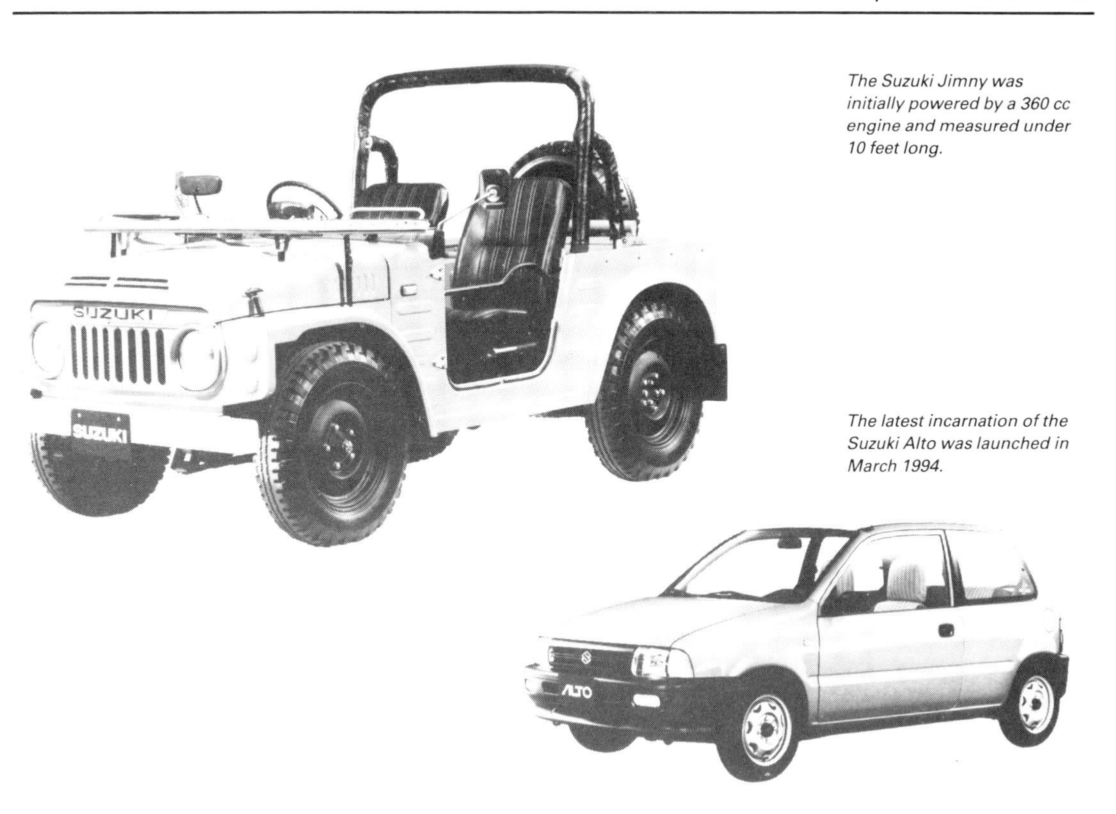

The Suzuki Jimny was initially powered by a 360 cc engine and measured under 10 feet long.

The latest incarnation of the Suzuki Alto was launched in March 1994.

MITSUBISHI

Mitsubishi's flirtation with microcars began in 1960 with the rather basic A10 500, a 493 cc rear-engined four-seater, which lasted until 1963. It was supplanted by the Minica 360 of 1962, a far more modern car conforming to the *Kei-jidosha* regulations. It used a two-stroke air-cooled twin engine of 359 cc. Its meagre 18 bhp gave it a maximum speed of 55 mph. Both saloon and estate versions were available. There was also a 600 cc non-'K' class car called the Colt 600 which also appeared in 1962.

Mitsubishi's replacement for the Minica was the new Minica of 1969, offered in saloon, estate and coupé versions. The latter initiated Mitsubishi's penchant for inexplicable names by being badged as the Skipper. Two engines were available: 359 cc air-cooled (30 bhp) and 359 cc water-cooled (34 or 38 bhp).

The model had a short lifespan, being replaced by the next Minica in 1972. This had only the water-cooled 359 cc engine, sufficient for a top speed of 71 mph. There was just one model, the two-door F4, although it was eventually replaced by the Minica Ami 55 with the same body but a 546 cc 31 bhp engine.

In the Japanese world of two-year production cycles, it is amazing that the third generation Minica lasted 12 years, when it was replaced by the new Minica in 1984. The same 546 cc engine now drove the front wheels and a turbo version and four-wheel drive were offered.

The next version of the Minica series arrived in 1989 (for export it was called the Towny). This had a new 548 cc three-cylinder engine with 29 bhp on tap (75 mph). With the relaxation of 'K' class engine size regulations, all Minicas received 657 cc three-cylinder engines with power outputs from 40 bhp up to a version with fuel injection, turbo and multi-valves with a power output of 63 bhp and a top speed of 93 mph, in which version it was known as the Minica Dangan ZZ. Another version came with one door on the driver's side and two on the nearside and was cryptically dubbed the Minica Lettuce. Still another version had an elevated roof rather like an ice cream van, measuring some 69 inches high. It was quaintly entitled the Toppo.

The latest Minica was launched in September 1993. The evolution was subtle, but the new Minica was distinctive for its high roof style. Both three- and four-cylinder engines were available, the latter going up to 63 bhp in the Dangan 4 version, which boasted four-wheel-drive, ABS and power-steering as standard. A new Toppo model was also available.

The coupé version of Mitsubishi's 1969 Minica was called the Skipper: this top-of-the-range Skipper GT could do 75 mph.

The 1984 Mitsubishi Minica Econo had a 546 cc engine driving the front wheels.

With an aerofoil on the roof, alloy wheels and chunky bonnet scoop, the Minica Dangan hardly looked like it had a mere 660 cc engine.

HONDA

Honda was an insignificant player on the Japanese car market until 1966. Before then, it had only the small S500 and S600 sports cars (not in fact 'K' class), but its fortunes were at once transformed by the introduction of the N360 in 1966.

The Honda S600 sports car – Honda's first car – was not a 'K' car, but it was very small by any standards.

This was a brilliant car. While most 'K' class cars were still powered by crude two-stroke engines sited at the rear, the N360 used a four-stroke 354 cc two-cylinder engine up front, driving the front wheels. Revving right up to 8,500 rpm, its power output was 31 bhp and a top speed of 71 mph was claimed. It looked modern, too, and handled very well.

The N360 was the first 'K' class car to be offered with automatic transmission (3-speed) in 1967. Export markets got the N500 and later the N400. But it was the N600 of 1968 which was Honda's pinnacle: its light alloy 599 cc engine developed 45 bhp and propelled it to a top speed of over 80 mph.

A development of the N600 was the bold Z Coupé of 1970. Idiosyncratic in style, it was a four-seater with a strange 'TV screen' hatchback. In Japan, it was only ever sold with the 360 cc engine, but most export markets got a detuned N600 engine (36 bhp) and the model was known as the Z600. It had a top speed of 75 mph and was notable for its high equipment levels, including an aircraft-style overhead console in the cockpit.

The brilliant N360 from Honda was sophisticated and fast. It was the first 'K' car to be exported to Britain.

Unusual in most respects, the Honda Z Coupé was only ever sold in the UK in orange with black stripes!

Another development was the Life of 1971. This was a two, three or four-door saloon with a 356 cc twin-cylinder engine developing either 21 or 30 bhp. In the case of the former, it was hardly a performer like the N360, having a top speed of just 56 mph. There was one further micro Honda, the Vamos, a curious doorless jeep with four seats and a pick-up bed. It used the N360's engine placed in the back.

Both the N360 and Z were withdrawn in 1974 in the face of falling 'K' class sales. Although Honda continued to make micro vans (like the TN7), it was not to rejoin the 'K' class arena until 1985, when its Today revolutionised what buyers could expect from a miniature car, just as the N360 had done nearly 20 years before.

The Today was the most sophisticated 'K' car available — and was also the most expensive. Its one-box styling was typically smooth and integrated and the cabin was spacious and airy. Under the short bonnet sat a 546 cc two-cylinder engine with 31 bhp on tap, for a top speed of 75 mph.

In 1988, the Today was restyled front and rear and got an all-new 547 cc three-cylinder engine with 36 or 42 bhp and a top speed of up to 90 mph. In line with the revised 'K' class laws, Honda expanded this to 656 cc and began to offer such options as four-wheel drive and fuel injection.

In late 1992, the Today's sophistication was restated with the introduction of an all-new Today with an unusual notchback body style.

Back-tracking slightly, in 1991 came the highly adventurous Beat, the first truly sporting 'K' class car. The formula was a mid-engined open two-seater sharing the minute dimensions of other 'K' cars but offering tremendous character and, above all, fun. The 656 cc three-cylinder engine was derived from the Today but incorporated variable tune intake ports for a peak output of 63 bhp at 8,100 rpm. 0-60 mph came up in 9.8 seconds.

Despite its tumultuous reception by the press (the author included), the Beat failed to live up to Honda's sales targets and production was suspended in mid-1992.

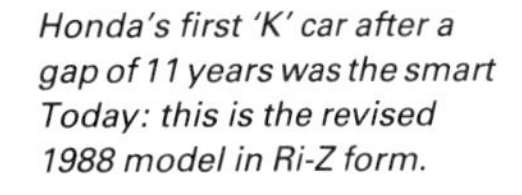

Honda's first 'K' car after a gap of 11 years was the smart Today: this is the revised 1988 model in Ri-Z form.

The tiny Beat was the first-ever mid-engined 'K' car and certainly the most hi-tech ever built.

DAIHATSU

Daihatsu began making small cars in 1955 when it introduced the Bee. This three-wheeled four-seater used a 540 cc air-cooled twin developing 13.5 bhp. This was followed by the Midget and Tri-Mobile, three-wheeled utility machines with 250 cc single-cylinder engines.

Daihatsu brought out its first 'K' car in 1960 when it introduced the Hi-Jet. This was a 360 cc pick-up which could also be bought as an estate car.

But the firm's first true passenger 'K' car came in 1966 when the Fellow offered buyers a typical brew of a 356 cc two-stroke twin engine and saloon or estate bodies. It was replaced by the Fellow Max in 1970, a rather better car also available as a coupé. It retained the 356 cc engine but in various states of tune up to 40 bhp (top speed 75 mph).

In 1976, the Cuore replaced it. It used a new 547 cc four-stroke engine providing 28 bhp and a top speed of 68 mph. In turn, its replacement, the new Cuore of 1980, used a more powerful 31 bhp version of the same engine, although in basic form, the old 28 bhp engine was available in a version dubbed the Mira. Turbo and four-wheel drive versions followed. For some export markets, a 617 cc engine was fitted, but not for the UK, where the Cuore was renamed the Domino and was sold from 1981 as one of the most economical cars on the market.

The third Cuore/Mira series arrived in 1985. In Japan, it came with a 548 cc engine but most export markets got the 847 cc unit (including Britain, where the car retained the Domino name). Both three and five-door versions were available. This became not only the best-selling 'K' car in Japan, but the best-selling car of *any* type, when it sold an incredible 300,000 units in 1990.

Daihatsu's Fellow of 1966 marked its arrival as a serious 'K' car contender.

The new front-drive Fellow Max (1970) was available with saloon, estate and coupé bodywork (saloon pictured).

For the 1990s, Daihatsu's 'K' range mushroomed: as well as the Cuore and Mira, there was the Leeza, a 2+2 coupé based on a shortened Cuore floorpan, which was also available in convertible Spider form. And secondly, there was the Opti, a voluptuously curved three-door hatchback. All shared the same 659 cc three-cylinder engine producing between 40 and 63 bhp.

After the Leeza was dropped, Daihatsu rejuvenated its 'K' class range in 1994 with a brand new Mira/Cuore — a more rounded, yet unadventurous, effort which was bound to repeat the sales success of its predecessor.

Daihatsu's Cuore/Mira series of 1980 was marketed in the UK from 1981 as the Domino, still with a 547 cc engine.

The world's most popular microcar: the Mira as brought into the UK by Daihatsu from 1993: 847 cc, five doors.

SUBARU

Subaru's 'K' car was the oddest of them all. It was an early contender, debuting in 1958, and was known simply as the 360. Its specification of a 16 bhp 356 cc rear-mounted two-stroke two-cylinder engine was entirely conventional, but its appearance was not: it looked like it had come from a jelly mould and sported the most curious 'eyebrows' over its headlamps.

Perhaps even more outlandish was the estate version, known as the Custom. A later version of the saloon, known as the K212 or Maia, had a more powerful (22 bhp) 422 cc engine.

By comparison, the 360's replacement, the R-2 of 1970, looked disappointingly normal, although it was more powerful (up to 36 bhp). Subaru's Rex of 1972 was another oddly styled beast initially using the old 356 cc engine. A year later, a new 31 bhp 358 cc unit was transplanted, still in the rear, for a top speed of 68 mph. Larger 490 cc and 544 cc engines followed.

The new Rex of 1981 very much resembled the current Suzuki Alto, but used the 544 cc engine up front (later a 665 cc unit). In 1988, the body style was again renewed. The following year, a new range of four-cylinder engines (547 cc and 758 cc) replaced the old two-cylinder unit.

To keep up with the times, the Rex received a 658 cc four-cylinder engine in 1990, including a 63 bhp supercharged version. Permanent and selectable four-wheel-drives were also on offer.

In 1992, Subaru attempted to elevate itself above its position as the 'poor relation' of Japanese 'K' car makers with the all-new Vivio. The engine remained the same but safety and accommodation were higher on the list of priorities. Initially two colours were offered: white or lilac. The Vivio was launched in Britain in 1992 as a rare example of a true 'K' class car sold in an export market. The novel four-wheel-drive version was joined in the UK by a front-wheel-drive model in 1994.

Subaru launched the most characterful, if not downright weird, 'K' car with the 360, later offered with a 422 cc engine.

The Custom (estate) version of the 360 was possibly even odder: strange curvy bits abound.

Subaru's 'K' car for the '80s was the Rex, available with a supercharger, as pictured here.

CHAPTER FIVE

THE NEW CLASS OF BUBBLES

JUST when everyone thought the microcar was dead and buried, when the micro lights of Europe had been thoroughly extinguished, a funny thing happened. The microcar came back.

In the early 1970s, there was really no microcar market at all. The birthplace of modern micros, Germany, had well and truly turned its back on the austere reminders of its past, the crude machines born in the days before it became Europe's most affluent nation. Britain, too, had largely waved goodbye to the breed and three-wheeler drivers had only the Reliant Regal and Bond Bug to tempt them.

But in France, the ground was exactly right for a spectacular rebirth of the true microcar. Just like most of the rest of Europe, France had laws allowing more free use of transport with small engines. The French Government said that anyone over 14 could drive a moped-engined vehicle (of up to 50 cc), *without* the need for a driving licence or *matriculation* (the French version of the MoT). As a further concession, the vehicle didn't need to carry a registration plate and so didn't have to pay for a parking space.

But with the death of the bubble car in the early 1960s, no-one really wanted to buy them any more. Quite why there should have been a sudden demand for such *voitures de ville* from the early 1970s has never been properly explained. The fuel crisis of 1973 may have been an important catalyst, but the ensuing boom outlasted the days of high petrol prices and economic fluctuations. Indeed, the boom is still resounding today.

A whole generation of strange and wonderful devices sprang up. They were all very small — typically around 7 feet (2 m) long — usually made from glassfibre and always rather bizarrely styled. Few could deny that these early French city cars were very crude, their little 50 cc single-cylinder engines buzzing headily to reach a government-restricted top speed of 45 km/h (27 mph). If you passed the equivalent of a Highway Code test, you could drive one with a 125 cc engine and reach the dizzy heights of 70 km/h.

But to the people who bought them, that didn't matter. For them, the microcar marked a return to the no-nonsense days of transport: times before Type Approval, seat belt regulations and paperwork. Mostly, they were an instant hit with the elderly who had never passed a driving test — they bought a microcar and never needed to: they simply got in and found instant freedom to travel around. Then there were youngsters, who could drive one of these *voiturettes* from the age of 14.

They lent themselves well to city traffic: very small, economical (typically around 100 mpg was normal), free from parking restrictions and easy to drive (automatic transmission was standard on most). But rural communities also saw the benefits of driving *sans permis*. There were also cases of drunk Parisian businessmen weaving their way back to work in these cars, safe in the knowledge that they couldn't lose their driving licences!

Soon, there were literally dozens of manufacturers across France churning out these tiny conveyances. As had been the case with cyclecars some 50 years before, there was a local manufacturer for most regions in France. People from that region generally bought that particular model: so people in Angers would probably buy a Mini-Comtesse, a Parisien might be tempted by a Vitrex Riboud and someone from Lyon would go for an Arola.

The first 'new breed' of French microcars were in fact electric (see Chapter 6). These began appearing at the end of the 1960s offering the very minimum in transport for very short trips. The Voiture Electronique arrived in 1968 and could be said to be the simplest car ever made. Teilhol, a large manufacturer which made the Renault Rodeo jeep, introduced its first electric car in 1972: the Citadine was not just the same size as a bubble car, it was the same shape, too.

With a legislative flexibility which most Europeans would find baffling, in 1976 the French authorities agreed to Guy Duport's request that the smoky, unreliable 50 cc moped engines in micocars might be replaced by larger-capacity diesels, to be treated the same as a moped engine. The agreement allowed for the fitment of diesels of up to 4 kW (nearly 6 bhp) with no change to the existing *sans permis* laws. This opened the door for a whole new breed of diesel *voiturettes* which, within a few years, became standard fitment for all microcars in France.

By 1975, there were half-a-dozen manufacturers; by 1980, that figure had more than doubled. In the heady boom of the early 1980s, there were as many as 30 makers of *voiturettes* in France offering over 50 different models.

The numbers of these cars being sold reached massive proportions, bigger even than the microcar boom of the 1950s. Ligier sold almost 7,000 cars a year at the height of the *voiturette* boom. Some of the biggest current French micro makers, like Aixam and Microcar, claimed to be selling over 4,000 cars a year, according to their latest figures.

Roughly parallel to the French experience, the same phenomenon was happening in Italy. In addition to the plethora of Fiat 500 and 126 based cars which appeared throughout the 1960s and 1970s, there was a consistent strain of true microcars from the early 1970s onwards.

Once again, many of the early cars were electric: the Zagato Zele (1972) set a pattern and became one of the best-selling electric vehicles of all time. Slightly earlier, in 1967, the Lawil made its first appearance and can be said to have sired the entire European microcar movement of the 1970s: its formula of a small engine and ultra-compact, basic bodywork was taken up by many other Italian manufacturers and also the French: probably the first modern petrol-engined microcar in France was a licensed version of the Lawil, called the Willam. Many Italian micros were eventually sold in France under different names.

But microcars never took off in Italy in the same way as in France. Older people did buy them in significant numbers, but never enough to support the huge diversity witnessed in France.

There were a few attempts to launch updated microcars in Germany and Britain but these countries' laws were not favourable to such cars: in Britain, a 16-year old could drive a three-wheeler with an engine of less than 50 cc, but it still had to pass an MoT test and Type Approval regulations and the driver had to have a licence. The situation was even tougher in Germany. So the two big markets for microcars in the 1950s were the least active in the 1970s and 1980s.

The only other significant activity in the world (apart from the Japanese 'K' class covered in Chapter 4) was in America. One could hardly imagine a more unlikely place for microcars to take root than the wide plains and chromed expanses of bodywork of the United States, but in reality there had been a long history of microcars in the 'States. The Crosley was perhaps the most successful, but there were plenty of microcar hopefuls throughout the 1960s.

Most American activity was in electric cars, where numerous firms offered shopping and commuting vehicles. Often these were nothing more than glorified golf carts, but the notion of pollution-free and economical transport struck a chord in congested America, especially California. Indeed, California has now legislated for zero-emissions vehicles to become mandatory by the turn of the century.

From the most basic *cyclomoteurs* of the 1970s, the typical French microcar gradually became more sophisiticated and correspondingly more expensive. The weird-looking contraptions with their suck-it-and-see layouts gradually gave way to cars which resembled larger cars in miniature, much like the cyclecar days and bubble car boom in decades past. Demand increased for cars with such luxuries as moulded bumpers, two-speed wipers, even leather seats and wood dashes. Many of these micros retailed for more than some cheaper models in the Renault range.

Today, the microcar's true home remains in France where there is still a healthy market. Recent legislation has forced microcars to carry registration plates, effectively removing one benefit of microcar ownership:

drivers have to find an allocated parking space for their cars just like any other vehicle. The wisdom of this when most *voiturettes* measure about 2.5 metres (under 7 feet) long is highly questionable.

FRANCE

WILLAM/LAWIL

M. Willam may be regarded as the father of the modern French microcar. As the president of Lambretta France, the importers of the Italian motor scooters, he was well placed to realise the potential of a microcar in France and, in 1967, presented his first car, named the Willam City. He collaborated with an Italian firm which undertook to make the cars and also marketed them in Italy under the name Lawil (*La*mbretta *Wil*lam).

The City was a very short (6 feet/190 cm) two-seater hardtop with a squarish body made of glassfibre and steel. The engine was a 123 cc single-cylinder scooter unit developing 5.6 bhp. Mounted at the front, it drove the rear wheels via a four-speed gearbox and could achieve a top speed of 43 mph (70 km/h).

This was only the beginning for M. Willam, who proceeded to form a microcar empire in France. In 1968 he introduced a convertible version of the City which he called the Farmer and four years later displayed a 125 cc prototype, which unfortunately came to nought. Then he introduced the Willam 500, which was an imported Fiat 500-based Italian microcar delightfully called the Baldi Frog in its home country, where it was also sold with 125 cc, 302 cc and 595 cc engines.

This was the start of a fruitful liaison with other Italian microcar constructors: eventually M. Willam would import cars from BMA, Casalini, Zagato and the San Marinese firm DECSA, all of which were marketed as Willam Lambrettas in France.

But the Willam continued to be developed: the Farmer 2 arrived in 1975 and there were long-wheelbase estate and pick-up variants. Around 2,000 cars per year were being sold at this time. There were even exports to Britain from 1969 to 1974, where the well-known conversion specialist Crayford sold the model. In 1978, Willam marketed a 49 cc three-wheeler with the name Cyclo, which was basically a modified Acoma Mini-Comtesse (see page 103).

The Lawil Varzina (aka Willam Farmer) sits in front of the Lawil/Willam City; engines were 123 cc scooter units.

A rare example of a Willam Farmer imported to the UK by Crayford: doorless body and the most basic equipment.

The Lawil-produced microcars were made up until 1986, by which time they were becoming rather long-in-the-tooth and uncompetitive. M. Willam depended on other marques to sustain his microcar empire.

Lawil itself also made some independent developments in Italy. In Italy, the convertible was always known as the Varzina and both it and the City were fitted with a 246 cc two-cylinder two-stroke 12 bhp engine in Italy. The Lawil Log of 1973 was a more rounded, modern microcar with an all-plastic body and bug-eye headlamps. Styled by Paolo Pasquini (later to make his own microcar called the Pasquini Valentine), the Log was listed for many years but never reached proper production. And Lawil produced a shameless copy of the All Cars Charly in the 1980s under the name Berlina A4, fitted with Lawil's familiar 246 cc two-stroke engine.

The Willam 500 was an imported Italian microcar delightfully called the Baldi Frog; it had a Fiat 500 engine.

The electric Citadine from Teilhol looked like the Isetta had come through a time-warp to 1972.

TEILHOL

Teilhol was a very early *debutante* in the French scene. It had been, since 1970, the contractor which built the Rodeo 'jeeps' for Renault and branched out in 1972 with the construction of the Citadine electric car.

To seasoned eyes, this might have been seen as the return of the Isetta. The shape was more angular but the profile was distinctively bubble car-like. There was even a front-opening door, albeit hinged at the top. A hatchback allowed entry to a rear parcel area.

The Citadine used a GRP body on top of a tubular chassis. Propulsion came from a 48v electric motor driving the single rear wheel. A top speed of 31 mph (50 km/h) was quoted and the range could be as high as 60 miles (100 km). Despite its tiny size (only 7 feet or 212 cm long), the weight of the batteries made it very heavy: all of half a ton (500 kg).

Other electric models followed: a 'pick-up' version was called the Citacom; the Messagette was a strange leisure car with a frilly canopy (later also a commercial); and the Handicar was intended for wheelchair-bound drivers.

By the turn of the decade, Teilhol was to become absorbed by the petrol-engined *voiturette* fever sweeping France. In 1981, it launched the Simply (later called the T50 and T125), a chunky convertible four-wheeler with 49 cc and 125 cc engines (later also a 325 cc diesel option).

This eventually became the basis of Teilhol's central model, soon offered with just 325 cc and 400 cc diesel engines and estate-type bodywork.

By the mid-1980s, Teilhol had abandoned its own electric cars (although it still offered an electric version of the Citroën C15 pick-up). In 1987, it introduced its own 'replacement' for the Citroën Mehari/Renault Rodeo, both of which were dead by that time. Called the Tangara, it used a Citroën 2CV6 floorpan with a pretty open jeep body and was also available with Voisin four-wheel-drive.

A new generation of Teilhols was born in 1989 with the individual-looking TXA series, offered with 325 cc and 505 cc diesel engines. These were rather up-market offerings, priced as highly as 78,000F (when the top-of-the-range Aixam was 66,900F). However, Teilhol was liquidated in spring 1990, only to be revived by two ex-employees who formed APP to market the TXA. This was a short-lived enterprise, however, and was taken over in 1992 by a new enterprise, Centuri, which lasted less than two years.

In the 1980s, Teilhol diverted its attentions to diesel-engined microcars. This is a Break 325TD.

The Teilhol range was latterly offered by Centuri Automobiles; the top-of-the-range model was known as the Monte Carlo!

CEDRE

The Toulouse-based engineer Francois Guerbet was another early microcar maker opting for electric power. His CEDRE Midinette appeared in 1975, a very compact car powered by a 1200-watt electric engine (a smaller 600-watt microcar called the Mini 1 (and later the Soubrette), had made its first appearance in 1974).

The CEDRE's angular style was pretty much without style; CEDRE described it as a "sheltered auto-cyle". It was a single-seater three-wheeler with a maximum speed of 31 mph (50 km/h) and maximum range of 37 miles (60 km). It had no pretensions to be anything other than the most basic transport. The definitive version arrived in 1978, sporting odd Z-back lines and a sliding Perspex door.

Virtually alone in the microcar world, Guerbet stuck to his electric guns, offering (slightly) more sophisticated versions of the basic idea into the 1980s. There was the bizarre 5x5 Solaire (which really did have five wheels!) and the Selectric kit for converting bikes to electric power (who says Sir Clive Sinclair was first?). However, by 1987 the appeal of the CEDRE had been outrun by changing times and the marque disappeared.

The electric-powered CEDRE was easily distinguishable by its sliding Perspex doors.

Last of many offerings from Vitrex Industries was the Garbo, powered by a 49 cc Peugeot moped engine.

VITREX

In its short life, Vitrex Industries of Paris was a prolific developer of microcars. It began its activities in 1974 with the Riboud, a very basic micro designed by an architect called Jacques Riboud and created by a French buggy manufacturer, Marland.

The Riboud used a 47 cc 2.4 bhp Sachs engine in a rather plain two-seater open body. Unusually, it was offered in both three- and four-wheeled forms. It filled a need, though, and was the cheapest *voiturette* around at 6,430F (in 1979). And it sold well: 560 in 1974 and 530 the following year.

Also in 1974 came the Addax. This was actually made by a different firm (ECAM) in Chambly and was even more spartan than the Riboud. A three-wheeled two-seater, the Addax used either a 47 cc or 50 cc engine with an automatic and three-speed manual gearbox respectively. There were also *Bord de Mer* and even *Sport* versions!

Vitrex mimicked Willam in importing Italian micros to France with the All Cars Snuggy and Snuggy Tobrouk. Its final fling was the Garbo, a snub-nosed little two-seater, this time with closed bodywork. Its 49 cc Peugeot monocylinder engine in the back produced 4 bhp. Vitrex stopped making cars just into the 1980s. Only one model, the Garbo, was snapped up by another maker: it was renamed the Puma and made until 1982.

MARDEN

Ten years before Renault launched its people carrier, the little Neuilly-based firm of Marden was making its own Espace: not a big MPV, but a tiny 82 inch (209 cm) long microcar. It straddled the electric and petrol ages with versions powered by an electric motor and a 47 cc Sachs petrol engine. Its mature-looking three-door bodywork was available in open and closed forms and it was well-finished and luxuriously equipped — quite a novel approach for 1975.

An attempt was made to power the car with a 435 cc Citroën 2CV4 engine in 1979, but this never got off the ground; if people wanted more power, they had to make do with the 124 cc version with 7.2 bhp on tap — top speed 38 mph (60 km/h).

Two changes of ownership in 1979 and 1982 hardly affected production which was running at a healthy 1,000 examples in 1981. New for 1983 was the strangely-named Marden Fetta, with new bodywork, four disc brakes and a front (as opposed to rear) mounted 50 cc Motobecane engine, later offered with 124 cc petrol and 325 cc or 400 cc diesels (the latter known as the Extra).

In the very utilitarian early days of the voiturette, the Marden was a pioneer of luxurious specification.

In 1987 came Marden's elegant Channel, a diesel engined front-wheel-drive *voiturette*. The following year's Alize was another smoothie based on the mechanicals of the Channel, alongside which it was sold until 1991, when it became Marden's sole model. As sales fell (from 860 in 1987 to 500 in 1990), Marden ceased to take such an important role in the marketplace and ducked out in 1992 after a creditable 17 years of production.

How to make a three-wheeled moped look chic: the cute Arola of 1976.

AROLA

One of the most successful of all French microcars was the Arola. It virtually created the form of future *voiturettes*.

First presented in January 1976, the Arola from Lyon was the brainchild of Daniel Manon. It was a very simple glassfibre-bodied design with three wheels. Its single-cylinder air-cooled two-stroke Sachs engine of just 47 cc capacity was typical of the genre. Producing only 3 bhp, it was not even powerful enough to propel the 242 lb (110 kg) car to the permitted top speed of 45 km/h (27 mph).

A four-wheeled version with a Motobecane 50 cc engine soon followed and there was also a pick-up called the Super Pratique (SP).

Like most *voiturettes*, the Arola gradually grew up. It became longer, more substantial, received independent suspension all round and was offered with the choice of a BCB 125 cc engine.

The rounded Minoto of 1982 was a model rescued from another microcar maker, Bel Motors, but it was not enough to save Arola, now based in Aix-les-Bains, from going bust. From a healthy 360 sales per month in 1978, the profusion of rival microcar makers forced Arola's production rate down to just 175 per month in its final year. Arola bowed out in 1983, although some of its expertise was rekindled by Aixam (see page 112).

LIGIER

Ligier is a name more commonly associated with Formula One motor racing and road-going sports cars. But from 1980, the creator of some of the most powerful cars ever made also produced one of the smallest and least ferocious cars of all time.

By 1981, the Arola had grown up to be a four-wheeler with such opulent standard items as opening windows and bumpers.

From Formula One to Formula Just Above Zero: the Ligier JS4 was nevertheless something of a design pinnacle in the French microcar world.

Guy Ligier's first foray into microcars was the JS4, a squarish but handsome two-seater powered by a 50 cc Motobecane engine driving the rear wheels. Alongside only a handful of other microcars such as the KVS, it had a metal body instead of glassfibre. It was, in its day, by far the most sophisticated of the French microcars, having four-wheel independent suspension, quad headlamps and large carrying capacity. For these reasons, the Ligier became a best-seller: no less than 6,941 were sold in 1981.

From 1982, a 125 cc engined version called the JS8 joined the range. Petrol engines were entirely supplanted by diesels from 1985, when the 330 (later called the Serie 5) was introduced.

With the Serie 7 of 1987, Ligier reflected the growing trend for microcars to become larger and more 'car-like'. The model was no longer a *monocorps*, but a two-box design just over 20 inches (52 cm) longer than the old JS series. Importantly, Ligier was the last French micro maker to switch from steel to glassfibre bodywork. Using the same 327 cc Lombardini diesel engine (later a 654 cc diesel), it could still be driven without a licence.

Production in 1987 had fallen to 2,500 units; in 1989 to 2,211, reflecting the more competitive market. Ligier's new model, the Optima (which arrived in 1989) eventually replaced the Serie 7. It used a single-cylinder 265 cc Robin diesel engine. By now, the specifications were becoming quite opulent with the options including tinted glass, rear fog lamps and even a leather interior! An estate-type version, the Optimax, was also available alongside a panel van equivalent.

With its markets expanding into Germany and Switzerland and its status as one of the best of the French micros, Ligier announced new 500 cc diesel and electric versions in 1992. Indeed, a Ligier won the annual 'Grand Prix' of electric cars, the Swiss Tour de Sol, in 1992!

In accord with French licensing laws a four-seater *avec permis* version of the Optima was also made available with a 617 cc three-cylinder engine — this was called the Optima 4. The latest chapter in the Ligier story was the preview at the 1994 Paris Salon of Ligier's important new model, due to go into production in Spring 1995. It had a 505 cc twin-pot petrol engine capable of taking it to nearly 75 mph.

In 1986, a British firm called City Wheels converted a batch of Ligier JS8s over to electric power.

Ligier's Optima opened up new avenues of sophistication and was one of France's best microcars.

ACOMA

This Angers-based enterprise was born as early as 1975 and quickly took 30% of the French microcar market, selling 3,000 cars a year by 1982. Its success was based on the Mini-Comtesse, a typically idiosyncratic looking single-seat thing with a 50 cc Motobecane engine mounted in the front of its polyester bodywork. It had a single front wheel and an odd folding gullwing door. The magazine *L'Auto-Journal* tested one in 1976 and concluded that it was so unstable the authorities ought to do something about it!

Four wheels arrived in 1978 with the new Mini-Comtesse, which looked somewhat like a Dalek from *Dr Who*. The Break version of 1979 (equally strange in appearance) had a large opening tailgate and an expansive glass area, although it was still tiny (measuring only 73 inches, or 185 cm, long).

For the 1980s, Acoma launched a series of more conventional microcars: the Comtesse Super Coupé, Comtesse Super Sport, Star and Starlette. Despite their grand names, they all had the same 50 cc engine (except the Star which used a 125 cc Sachs). For a while, Acoma rated as one of the big microcar players, but its fortunes began to wane in the face of fresh competition. Production had come to a halt by 1984.

The 1978 Mini-Comtesse made by Acoma was a dumpy little microcar which resembled somewhat a Dalek.

The Comtesse Super Coupé was Acoma's bid for normality. The more typically weird Mini-Comtesse Break can be seen in the background.

ERAD

The name Erad is one of the most respected in French microcar circles and, at the time of writing, it is the oldest surviving producer of *voiturettes*.

Erad was formed in 1975 (its name was an acronym of Etudes et Realisations du Douaisis). Its first model, the Capucine, was a cute little single-seater with a body made from a plastic called Altuglass. The usual 47 cc Sachs 3 bhp engine sat in the rear end. Just about the only technical novelty was McPherson strut front suspension.

But it proved very popular, selling 650 examples in its first year. Its bodywork matured for the coming decade and Erad, like most French firms, offered a diesel alternative from 1981: in this case, the Farymann 290 cc single-cylinder (which had achieved the world record for consumption at no less than 1,919 mpg!). From this date, the petrol-engined model was also available with a 123 cc BCB engine. Erad also introduced a small truck called the Utilit.

Perhaps Erad's most endearing model was the Midget of 1982. As its name suggested, it was an accurate replica of the 1936 MG Midget. One big difference: it was only 9 feet (270 cm) long! And its engine was no bigger than 125 cc. It weighed just 700 lbs (320 kg) but was a fair performer. Costing 50,000F in 1984, it was expensive as a *voiturette* but cheap as a replica. Later versions used a three-cylinder Kubota 600 cc engine.

The Capucine got a new body shape for 1984, although it retained its compact dimensions (only 77.5 inches or 197 cm long) and had the same mechanical basis. A convertible version, dubbed the Plein Air, was also introduced at this time and an extended wheelbase four-seater arrived for 1987.

Erad's up-market model in 1986 was the new 6.50 D series, which came with a choice of 290 cc, 304 cc and 400 cc diesel engines and had more attractive bodywork.

The Capucine finally gave way to a new model, the Junior, in 1988. Designed by the same man responsible for most of Erad's output, Benoit Contreau, the Junior was a pretty two-seater with a large glass area, available in open and closed forms and with 304 cc or 400 cc diesel engines. It slotted in at the base of the range with a price which was the cheapest of any *voiturette* of the time: just 35,000F. Its main suprise was that the whole glass canopy tilted forward for entry.

A nine-foot-long replica of a pre-war MG Midget with a 125cc engine? The Erad Midget could only come from France.

Erad's impressively modern offering for 1986 was the diesel-engined 6.50D.

Despite its budget price, Erad's Junior succeeded in offering a clean, stylish look. Entry was via a hinged canopy.

Quite a stir was caused when Erad launched its Spacia in 1990. This was a *monocorps* design with very attractive lines. A choice of diesel engines up to 505 cc was offered and performance from the most powerful (for which you needed to pass your driving test) was very good for a microcar: up to 47 mph (75 mk/h). Like the Junior, it was also available with an electric power plant.

By 1991, Erad had sold over 25,000 of its *voiturettes*, which made it one of the biggest French firms. An important newcomer in 1993 was the Agora. This marked a return to the ultra-basic theme, with the simplest two-seater construction, a canvas roof and extremely compact dimensions (just 210 cm, or 82.7 inches long). Erad called it a "2CV for the year 2000" and prices began from just 40,000F — the cheapest car on the French market.

Currently the unusually wide Erad range consists of the Agora (electric, diesel, A4 diesel); the 6.400 and 6.500; the Spacia (electric, diesel, 2 or 4 seats); the Midget; and the Utilit micro-truck range.

Years before the Renault Twingo, Erad offered the brilliant one-box Spacia, powered by a 505 cc diesel engine. To the left is the 1993 Agora – a return to the ultra-basic theme.

DUPORT

It was Guy Duport who was responsible for opening up the French microcar market to the extent it enjoyed in the 1980s. It was he who approached the French government with a proposal to change the law regarding the use of cars without a driving licence. As the law stood, only cars of up to 50 cc could be classed as useable *sans permis*, but in 1976 Duport managed to persuade the diesel-friendly authorities to allow diesel-engined microcars to be classed in the same category, so long as their engines developed no more than 4 kW (6 bhp). This was a historic move.

Guy Duport had made his fortune manufacturing teleskis but quickly cashed in on the new law in 1977 by branching out into microcars with his Duport Caddy. This was the first of the diesel-engined micros and held, for a while, the title of 'the world's smallest diesel car'.

Its engine was a rear-mounted 510 cc Lombardini diesel which developed from 6 to 10 bhp and it made clever use of Renault 4 steering and brakes. With a top speed of 47 mph (75 km/h) in the most powerful variant, it was also a bit of a tearaway among micros of the 1970s.

Duport's range for the 1980s was based around the same Caddy bodyshell and came with various engine options from 359 cc up to 602 cc (all diesels) and came with the designations 311, 411, 511 and 611.

There was also the rather more mature Parco, introduced in 1980, which used the Renault parts bin for certain areas of its bodywork. It had four seats and a longer body (up to 9 feet 6 inches, or 290 cm).

1986 saw a new range of *voiturettes* replace the old 511. The two-seater Parco 50 Diesel (and one year later, the longer 125/4) resembled the Parco but remained a *voiture sans permis*. As the Parco was discontinued, the range was streamlined to just three models: the 360A, the 400A and the 125/4.

Guy Duport was the first maker of diesel-engined voiturettes *in France. This is the four-seater Parco.*

Two new models arrived in 1990: the D50, the newest Duport *sans permis*, and the Onyx, a jeep-type microcar undeniably inspired by the Citroën Mehari using the same 325 cc diesel engine as the D50. The 125 model eventually got a 654 cc diesel engine and was even dubbed the GT!

Duport was never a really big seller in the cut-throat French market (he sold only 800 cars in 1989), but he was a consistent player. Guy Duport finally sold his interests to a new firm, SND (Societe Nouvelle Duport), in February 1992, but this enterprise finally closed its doors in 1994.

Duport's Parco 50 was a smaller two-seater hatchback which made clever use of the Renault parts bin.

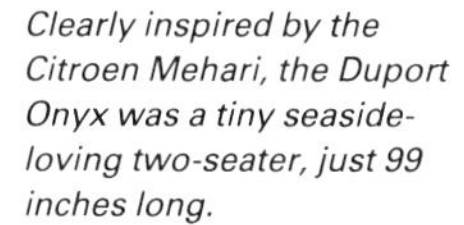

Clearly inspired by the Citroen Mehari, the Duport Onyx was a tiny seaside-loving two-seater, just 99 inches long.

FLIPPER

Somehow the name of this most unusual French microcar suited it perfectly. The Flipper, which was first shown in 1978, had one overriding feature: its narrow-set front wheels to which the 47 cc Sachs engine was attached could be turned through 360 degrees. This meant that an indicator had to be fitted in the car to prevent the driver shooting off in a violently unexpected direction!

Apart from its appearance, the Flipper was in most other respects fairly normal among French micros. It had an automatic gearbox, independent suspension and a top speed pared down to 25 mph (40 km/h). Its plastic bodywork *was* unusual: it was made in two halves, then bonded together to form a chassis-less monocoque shell.

It became more normal. The rotating power plant was replaced by standard-issue front-wheel drive in the Flipper II of 1980. That was only slightly redeemed by the name of the open version announced a year later: the Flipper Donky. The firm's last fling was the Flipper III, a disappointingly ordinary microcar with a 50 cc Polymecanique engine. The market was already overrun with such cars and the Flipper had flapped its last by 1984.

Idiosyncracy on wheels: the 1978 Flipper could turn in its own length thanks to its pivoting front wheels.

MICROCAR

Microcar's first offerings were the RJ 125 and DX 125. They established the firm's superiority in the French market.

From unpromising beginnings, the aptly-named Microcar became the best-selling French microcar of all time. This was largely due to the big-time backing it had: its parent company was Janneau, Europe's leading pleasure boat manufacturer.

Janneau bought the rights to a very basic three-wheeled microcar called the Mini-Cat in 1980. This was actually designed by racing car builder Serge Aziosmanoff in 1978. Janneau's plan was to move into the expanding *voiturette* market, utilizing its glassfibre skills. It modified the Mini-Cat into a four-wheeler, to be called the Microcar.

The first Microcar, the RJ49, was launched in 1980, a quite unexceptional little two-seater with a 49 cc engine. Its principal attraction was the price: at 21,400F it was just about the cheapest four-wheeled *voiturette* available. Its target market was the rural communities, with whom it immediately struck success. It was joined in 1981 by a slightly better-looking model, the DX49, distinguishable by its wrap-around front screen and vaguely 'coupé' styling. A convertible RJ (later known as the Solea) arrived in 1983, alongside 124 cc versions of both models. While the DX was not a great success, the RJ49 and RJ125 definitely were: indeed, many examples of the RJ can still be seen on French roads today.

For 1984, Microcar launched a new range of cars under the name 50. Equally bland in appearance, they were also available with diesel engines, which eventually became the mainstay of the range. By this time, Microcar was definitively France's leading brand of *voiturettes*. By 1987, it had sold its 20,000th *sans permis* car and annual production stood at about 3,000. Its range of engines continued to expand, too: there were now four options from 50 cc to 600 cc.

Why the Microcar scored such a huge success is probably due to a 'snowball' effect initiated by Janneau's huge distribution network of 190 concessionaires and 700 sales outlets through France, offering unrivalled after-sales service. In this respect, it was not like most other micros which were sold mostly in their own locality.

In 1988, just after the parent company was returned to French ownership (it had previously been in American hands), Microcar at last came up with a good-looking car which it called the Spid!, probably the only car ever to have been marketed with an exclamation mark in its name. Two diesel engines were offered: a Yanmar 273 cc (*sans permis*) or a Ruggerini 654 cc (*avec permis*), dubbed the 30S and 90S respectively.

It was a big hit. In its first full year, the Spid! sold 3,405 examples. An even better-looking version, occupying a position as a budget micro, was the Lyra of 1990, sold only with the smaller Yanmar diesel; the Spid! became an *avec permis* model only from this date. But Microcar lost its pole position to Aixam in 1990 (3,200 versus 4,000 sales).

In 1992, the Lyra became Microcar's sole model, now offered with 273 cc and 505 cc diesel engines at prices from 57,000F to 76,000F (no longer cheap and cheerful). The top versions even came with cigar lighters, side impact beams and metallic paint! Also in 1992, Microcar launched itself for the first time into the electric arena with a battery version of the Lyra (also known as the Light in some markets). So the French microcar scene came full circle, with a return to the electric days which began the 'new age' of the micro.

A new model announced in 1994 was the Newstreet, a project jointly created with La Prevention Routiere. Aimed at the 14-18 year-old market, it was a jolly-looking cabriolet powered by a Yamaha moped engine. Currently, annual production of all Microcar models is running at around the 3,000 mark.

DELSAUX

The Delsaux Modulo was typical of the budget end of the *sans permis* microcars. Introduced in 1980, it had a 47 cc Sachs engine mounted in the rear and very basic plastic bodywork. 200 were sold in 1981. In 1982, it was improved with revised bodywork and renamed the Modulo Minimax but, by 1983, Delsaux had disappeared.

Delsaux's Modulo of 1980 was cheap, pretty but not very successful, lasting only three years.

GMT

The Rivelaine was presented by Generale de Mecanique et Thermique (GMT) in 1981, initially with a Fichtel & Sachs 47 cc engine (later a 49 cc Motobecane unit). The four-wheeled two-seater was fairly typical of the breed but offered such luxuries as winding windows, locking doors and a heater. It had vanished by 1983.

The GMT Rivelaine was typical of the dozens of short-lived microcar makers which sprang up in the early 1980s.

TOMCAR

There was one thing the 1983 Tomcar had which few other French micros of the time could offer: it didn't look like anything else. Its bug-eye headlamps and separate wings hinted at a bygone era and had a definite sense of humour.

However, it shared all the usual microcar traits: a 49 cc Peugeot engine (later 125 cc petrol and 325 cc diesel options) compact dimensions (97 inches/246 cm long) and two seats. But it was ahead of its time in offering front-wheel-drive and disc brakes all round. Two body styles were offered: an enclosed *berline* and a cabriolet.

By 1985, the Tomcar had disappeared, although it did briefly resurface in Spain in 1986 as the Pypper.

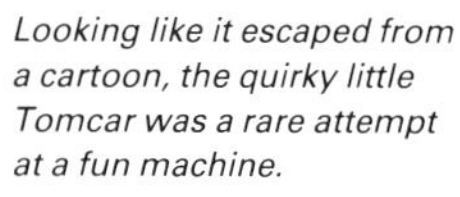

Looking like it escaped from a cartoon, the quirky little Tomcar was a rare attempt at a fun machine.

GATEAU

To name a car the Gateau (which translates as 'cake') — the parent firm was called Gateau International — may seem a trifle jocular, but then the firm's first model was rather tongue-in-cheekedly named the Egzo-7 (pronounced 'exocet')...

This appeared in 1983 as a rather handsome beast with one-box styling by Joel Bretecher. A big range of engines was offered, from 50 cc up to 500 cc and there was even an electric version. A model for handicapped drivers was on offer, too.

By the following year, the model had been renamed Egzo-3, one of many name-changes which included Maximini, Break and Grande. Gateau quickly abandoned its petrol-engined models to concentrate solely on diesels.

In 1988 came a new model, the slightly larger Vison with a 325 cc diesel engine which was the firm's mainstay until 1991, when the model was revised with a new front section and renamed the Forum. Gateau microcars had been withdrawn by 1992.

Gateau produced a range of modern voiturettes *from 1983. This is the Vison of 1988.*

Aixam was – and still is – the best-selling microcar brand in France. This is the A540 Twin of 1992.

AIXAM

Aixam was born out of the ashes of the trail-blazing Arola (see page 101), relaunching production in 1984 with the 325D diesel-engined *voiturette*. Aided by an already-strong customer base which remembered the Arola, Aixam was, within a year, the third largest manufacturer of this class of vehicle. To celebrate, it launched the faster 400D.

By 1986, the old Arola factory in Aix-les-Bains had become too small for the growing enterprise and a new 3,000 square metres factory was built. In 1987, with the launch of a new range of vehicles, Aixam displaced Microcar as the largest microcar maker, taking some 35% of the entire market. Its massive dealer network was partly responsible (today, it boasts an amazing 900 outlets in France!).

The new models were the 325i and 400i, available with two or four seats. In 1988 the 325i passed the 30 mph crash barrier tests (which the Government eventually decided not to impose on microcars). But Aixam's point had been proven — these were impressively tough little cars.

In 1989 arrived the three-cylinder 600i model, shortly followed by the 500i bi-cylinder. The 400i Twin of 1991 represented the first-ever water-cooled twin-cylinder *sans permis* microcar.

From 1992 a new range of models was created: the A540 and A550, the latter with larger bodywork and seating for four, aimed at the A4 class.

In 1993 everyone was stunned by Aixam's creation of a new marque, Mega, to produce the phenomenal Mercedes-Benz V12-engined, four-wheel-drive, Track supercar — the absolute antithesis of the little micro-cars. The price difference between the cheapest Aixam — at 50,900F (£6,400) and the Track (£240,000) tells all.

Aixam remains the largest maker of microcars in the world, having sold 3,617 of them in 1993 — some 1,179 better than nearest rival Microcar. A new budget version of the A540, dubbed the Record, looked set to cement Aixam's enviable record for the future.

SILAOS

Replicas of veteran cars (many of which were themselves very small) began as long ago as the 1950s with cars like the Rollsmobile (see Chapter 2). Among modern replicas, one of the most attractive, if impractical, was the Silaos Demoiselle, introduced in 1985.

Created by ex-Renault stylist Joel Michel (who worked on the R5 Turbo), it was an authentic replica of the Bugatti Type 56 electric car used to ferry customers around the Bugatti factory in the 1920s. From its large wire wheels to its wicker luggage basket, it was original and seductive.

As well as 47 cc petrol and 430 cc diesel engines, the firm which made the Demoiselle, Projet Plus of Dieppe, offered an electric version. It weighed only 430 lbs (195 kg) and cost about the same as other *voiturettes* (39,900F in 1986). By 1988, however, the Demoiselle had disappeared.

A successful attempt to produce a micro-replicar: the Silaos Demoiselle reproduced the lines of the Bugatti Type 56.

ITALY

SAVIO

Making jeeps (or, in Italian parlance, 'torpedos') using a Fiat 500 floor-pan was a favourite past-time of Italian coachbuilders in the 1960s and early 1970s; when the Fiat 126 came along, many adapted their cars to suit the new basis.

Some, like Savio, produced new designs. Its previous Jungla had used the mechanicals from the Seat 600 but the all-new replacement, also called the Jungla, used the mechanical basis of the 126. It was first shown at the Turin Show of 1974. A four-seater, it featured a folding wind-screen, doorless body and a soft-top. Savio sold the model in Italy until around 1983.

The Fiat 126-based Savio Jungla makes its début at the 1974 Turin Motor Show.

ALL CARS

The little Charly was first presented by a firm called Autozodiaco in November 1974. An amusing and genially shaped three-wheeler, it used a plastic two-seater body composed entirely of straight lines. Typical of Italian city cars, it used a Minarelli 49 cc engine.

A four-wheeled version, predictably called the Charly 4, arrived in 1975 and sported a rather more potent 125 cc Lambretta engine in an approximately similar body.

Production had transferred to a new enterprise called All Cars by 1978, which was affiliated to Automirage (see page 117). The model now used a 50 cc Morini 4.5 bhp engine and was renamed, brilliantly, the Snuggy. A new convertible version painted in obligatory khaki and carrying a spare wheel on the back was also introduced with the yet more inspired name of Snuggy Tobrouk. Later versions switched again to Motobecane 50 cc engines and were even offered with 250 cc units.

The design was so good that the Italian microcar grand-daddy, Lawil, made a more or less direct copy of it. But by 1985, All Cars was no more.

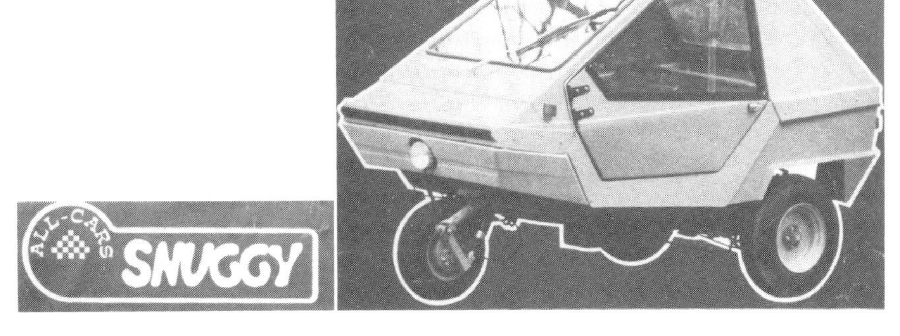

Somehow, the name All Cars Snuggy suits this tiny three-wheeler perfectly. It was presented as "the car of the future."

GIANNINI

Giannini is a name legendary in Italy for its tuning prodigies based on Fiats and is second only to Abarth for its illustriousness. It produced a whole string of tuning packages for the Fiat 500, from the mildest power hikes to full-blown racing engines, often involving boring the 499 cc engine out to as much as 694 cc and 41 bhp.

Giannini's first sports car since 1953 was the Sirio, first shown in 1972. Its extraordinary lines were created in glassfibre and the most striking feature of the car was its curved windscreen and detachable glass roof. It measured just 35.4 inches (90 cm) high to its rear spoiler, under which sat a Fiat 500 engine bored out to 652 cc and developing 35 bhp. Giannini claimed it would reach 99 mph (160 km/h). It is believed that the two-seater never entered production, although it was listed as a production model.

Giannini's souped-up Fiat 500s were built until 1975, supplanted by hot versions of the 126 and other Fiats right up to the present day (with the Cinquecento and Punto). In 1992, it announced that it would be making a limited number of replicas of its most potent 500 using original Fiats, to be called the 500 Corsa. Unfortunately, these were intended for track use only.

Giannini, the celebrated Fiat tuner, essayed this extraordinary Fiat 500-based two-seater in 1972: the Sirio.

The very compact 50 cc Sulky was produced by the Italian firm Casalini in three and four-wheeled form.

CASALINI

The Piacenza-based firm of Casalini was responsible for bringing into the world one of the best-named cars of all time: the Sulky. This appeared around 1975 as a pretty basic little two-seater powered by a 50 cc or 60 cc engine. Unusually, it had steel bodywork, which was all straight lines, mounted on a tubular steel chassis.

As the model developed(!), it got two headlamps instead of one, the option of four wheels (from 1980) — in which form it was known as the Bretta or David — a Break estate version and a larger engine option (125 cc). In the 1980s, Casalini was making as many as 1,000 cars a year, helped by strong demand through the French importer (Willam). Alongside BMA, Casalini's Sulky is, at the time of writing, the lonely remnant of the Italian microcar industry.

BMA

Symmetrical shapes which minimise tooling costs are not uncommon: the Champion and Dornier Delta of the 1950s are but two examples. However, there was no car quite as bizarrely symmetrical as the BMA Brio.

This was quite the most surreal offering from Italy, a country noted for its outlandish *carrozzerie*. Its polyester resin bodywork resembled a wedge of sculptured gorgonzola turned on its edge. The fully convertible version was hardly more normal.

Technically, the Brio recalled the days of ultra-basic transportation. Its tiny 47 cc Sachs engine sat in the tail, from where it drove the rear right wheel only. This putt-putted the Brio to a distinctly un-brio-like top speed of just 23 mph (37 km/h).

It had joined, in 1978, one of the more established Italian microcars, the BMA Amica, first seen in 1971. This was another three-wheeler, rather larger than the Brio, which sported engines from 50 cc up to 223 cc. Its plastic bodywork featured gullwing doors, no less.

BMA introduced its Nuova Amica in 1980. This had rather more normal-looking bodywork and three or four wheels, offered with a choice of 50, 125, 250 or diesel 360 cc engines.

BMA products were marketed in France by Willam throughout the 1970s and into the 1980s, which helped BMA sell around 500 cars per year. The Brio vanished in about 1986, leaving only the two Amica models to continue the line. They remain available at the time of writing, with 125 cc or 250 cc engine options.

This alien pod (left) is a BMA Brio, a bizarre device made from 1978 in Italy. Alongside it is a long-wheelbase Willam Break.

AUTOMIRAGE

Automirage was one of Italy's biggest beach buggy manufacturers, offering three models, the Mirage, Pirana and Moon. Its first venture into microcars was the Pick Wick of 1975, a rather basic buggy/jeep cross-over based on the Fiat 126.

Soon it had caught the bug and began to abandon buggies in favour of its microcars. The second model was the Mirage 3 of 1976, a glassfibre-bodied angular three-wheeler with a 50 cc Morini engine sited in the rear. This drove the rear wheel by chain through a three-speed gearbox. It was officially described as a '1+1'-seater. Later it received engines of 125 cc and 250 cc capacity.

In 1982, it was joined by two new models sharing similar bodywork, the S3R and S4R, denoting the number of wheels each car had. These were fitted with engines from 125 cc to 250 cc, had automatic transmissions and two seats. Automirage became nothing more than a mirage in 1985.

Automirage produced the very simple Mirage 3 from 1976. Like all moped-engined microcars in Italy, no driving licence was required to pilot it.

The Fioretti F50 was a brave but unsuccessful attempt to launch an open car based on a Piaggio truck chassis.

FIORETTI

The Rome-based firm Autofioretti introduced the glassfibre-bodied Firoetti F50 in 1978. It was based on the well-known three-wheeled Piaggio commercial with its 50 cc air-cooled single-cylinder engine. It was optimistically described as a 'targa-top' and did not last more than a couple of seasons.

DECSA

DECSA's Lisa cannot be excluded from a review of microcars as it was San Marino's first ever motor manufacturer, appearing in 1982. Unfortunately, that was just about the only exceptional thing about it. It had a normal-looking two-seater glassfibre body and a choice of 50 cc and 123 cc single-cylinder engines or a 250 cc twin. Both three- and four-wheeled versions were built. Helped by French sales, it continued in production until about 1987.

GREAT BRITAIN

MICRODOT

William Towns is a name made famous by his work for Aston Martin (he designed the DBS V8 and Lagonda). He always harboured a passion for city cars, the first expression of which was the Mini-based Minissima (initially called the Townscar) of 1973, which was even given to British Leyland for evaluation.

Towns's city car project for 1976 was the Microdot, which he described as "a bubble car for the 1980s". It was a very small car with a large glass area (incorporating glass gullwing doors) and seating for three abreast. Initially, the power source was electric. Towns found a backer for the project in the Bentley Mk VI sports car makers Mallalieu in 1980, who were to build a batch of cars with a petrol engine (Mini, Reliant and marine two-strokes were suggested).

To be offered in two versions, an economy and a luxury, at prices from £4,000 to £7,000, the Microdot would surely have been too expensive. As it was, Mallalieu's bankruptcy in 1981 put paid to the plan. The Microdot was eventually acquired by Daihatsu GB, which installed a 547 cc Domino (Mira) engine into it. This is currently owned by the Patrick Museum in the Midlands.

William Towns's compact three-seater Microdot was destined for production but remained a one-off.

BAMBY

A microcar enthusiast, Hull-based Alan Evans was inspired by his acquisition of a Peel P50 (see Chapter 2) to attempt to create a modern equivalent. In 1983 his Bamby made its bow.

Like the Peel, the original Bamby was a single-seater glassfibre-bodied three-wheeler with a 49 cc engine. It weighed all of 235 lbs (107 kg) and could return 100 mpg. And it looked pretty good, too, although its price of £1,597 was decidedly on the high side.

Soon some design changes were made: the single gullwing door was changed to a conventionally hinged door, there were twin headlamps in place of the original cyclops lamp and the air vents, which had let a wasp into one customer's car, were blocked off using a kitchen sieve. The engine was progressively upgraded from a 49 cc Minarelli to Yamaha, then Suzuki moped units.

An initial production rate of 20 per month proved over-optimistic and the Bamby died after a only year or so, as founder Alan Evans left the project to a businessman who liquidated the venture in a matter of months. Probably around 50 Bambys had been made.

CURSOR

Alan Hatswell's Replicar Ltd, purveyors of kit-form replicas of Bugattis, Jaguars and Ferraris, launched its 'revolutionary micro vehicle' in 1985. And a decidedly strange beast it was.

Called the Cursor, it was a single-seater three-wheeler whose main selling-point was that it was classed as a moped and could therefore be driven by 16 year-olds. Its extremely odd-looking glassfibre body sat on top of a tubular steel chassis. Cryptically described as a 'GT hatchback convertible', it boasted nothing more sporting than a 49 cc Suzuki CS50 moped engine; presumably the GT tag derived from the fact that it was mid-mounted. A top speed of 30 mph was offset by a fuel consumption of about 90 mpg, and all for £1,724.

After about 50 had been made, a two-seater version followed with a more powerful Suzuki CP50 engine and gullwing doors fitted. The final ten or so of the 50 two-seaters built had Honda Vision moped engines. Most of the later models were exported to Vienna, Austria.

Replicar claimed grandly of the Cursor that "the impact on driving will be much the same as when the Mini was launched in the late 1950s." In fact, drivers (even of the 16 year-old variety) hardly gave it a cursory glance. Its chances cannot have been helped by the appearance at the London Motorfair of a Cursor covered in gold metalflake paint! The project was sold on to a firm in Belgium, where production was set to commence at the time of writing.

The Cursor looked a bit like it should go home to a fairground, but it was driveable on British roads by 16-year olds.

PORTUGAL

ENTREPOSTO SADO

Virtually every European country has had a microcar maker. Portugal is no exception, although it took until 1982 to happen. The Entreposto Sado 550 was made in Lisbon and featured a two-seater glassfibre body on a steel tube chassis equipped with a Daihatsu 547 cc four-stroke two-cylinder engine. This made it a bit of a 'go-er' among microcars, listed as being capable of 69 mph (110 km/h). In other respects it was entirely conventional and offered city dwellers in Portugal the chance to beat the traffic until 1986.

SWITZERLAND

THAON

Swiss microcars have been pretty thin on the ground over the years, but there have been some notable exceptions. Sbarro's Carville (described in Chapter 6) is one; the Thaon AT2 is another strange addition.

The brainchild of Andre Thaon of Montreux-Clarens, the AT2 was an exceptionally small electric car first shown at the Geneva Motor Show of March 1975. It measured just 4 feet 11 inches (150 cm) long and 31½ inches (80 cm) wide and, in its lightest version, weighed a mere 232 lbs (106 kg).

Two versions were offered: one for use as a factory or security vehicle, the other for the road. The latter had more powerful motors enabling it reach a top speed of 37 mph (60 km/h) as against just 19 mph (30 km/h) for the 'special' version. Each front wheel was driven by a separate motor which turned with the wheel in an exceptionally narrow turning circle of just over 9 feet (2.9 m).

The two passengers sat in tandem but the only way to do this was to offset the rear seat so the occupants overlapped, which cannot have been particularly comfortable in so narrow a car. Entry was via sliding doors on either side.

The production life of the Thaon was very short and few, if any, cars were sold.

Portugal's interpretation of a modern microcar was the Entreposto Sado which used a 547 cc Daihatsu engine.

Made in Switzerland, the electric Thaon was one of the shortest and narrowest cars ever built.

GREECE

DIM

If ever there was an unfortunate name for a car, then DIM was it. It was actually a contraction of Dimitriadis, the name of the man who designed this car in Athens, Greece, and who had been responsible for building the Fuldamobil under licence in Greece. The DIM was basically a rebodied Fiat 126, with its own (not especially elegant) glassfibre body which seated four. It was 6 inches (15 cm) longer than a 126 but in most other respects duplicated its Italian forebear, with the exception of the Fiat's tendency to rust away. First shown at the 1977 Geneva Motor Show, it continued to be seen at international shows, but full-scale production never happened.

USA

TRI-PED

America's output of microcars in the 1970s was miniscule. One design which did make it into production was the 1979 Tri-Ped Microcar made in Farmingdale, NY.

The Microcar was little more than a three-wheeled moped, which exempted it from the stringent American emissions laws. The 50 cc two-stroke engine drove the rear wheels by chain. The specification was basic: brakes on only two wheels, the simplest tubular frame with zip-up plastic sheeting for weather protection and handlebar steering. The driver had to pedal until the centrifugal clutch fired up the engine.

Weighing only 160 lbs (73 kg), the owner was at least rewarded with remarkable fuel consumption: an average of 100 mpg was claimed.

INDIA

SUNRISE BADAL

In India, the emphasis has always been on affordable transport. No other car tried as hard to offer this as the Sunrise Badal. Made in Bangalore from 1978, it was a curious three-wheeled vehicle with an oddly-shaped glassfibre body whose most unusual feature was that it had two doors on the nearside but only one on the offside! The manufacturers inexplicably described it as "futuristic"!

Despite its size (over 10 feet or 308 cm long), it weighed only 882 lbs (400 kg). This meant the rear-mounted 198 cc single-cylinder 10 bhp engine could power it to a top speed of 47 mph (75 kh/h). There was room for four passengers, making it popular as a taxi.

In 1981, the Badal got a spruce-up in the design department and gained an extra wheel at the front. But this could not hide the fact that the Badal was basically a very crude and poorly-finished car and the model was withdrawn the same year in favour of licensed production of the rather larger British Reliant Kitten four-wheeler. This was known as the Sipani Dolphin and was virtually identical in all respects to the Kitten. This received development over the years, including a four-door version called the Montana and a deathly slow diesel model. These ex-Kittens never really got into their full production stride and were effectively replaced by Sipani's new project in 1994: licenced production of the Rover Montego.

The Indian-built Sunrise Badal was offered in both three and four-wheeled versions during its three-year life.

Fiat 126 basis and a glassfibre body for the Greek DIM 594.

A rare American microcar: the Tri-Ped Microcar, basically a moped with three wheels and rudimentary weather protection.

CHAPTER SIX

ELECTRICKERY

ELECTRIC cars were among the pioneers of the motor industry and at one time looked like they might give the internal combustion engine some competition. But they fluttered out, along with steam-powered cars, leaving camshafts and carbon to reign supreme.

The reason was the impracticality of electric power compared to fossil fuels. And that remains the essential stumbling block even today. Electric cars are a great idea waiting for the right technology to redeem them. With battery technology as it stands, lead-acid batteries are excessively heavy and the newer silver-based cells are very expensive and need replacing at regular intervals. Range is limited between charge-ups and performance is typically very lacking.

That's the down side. To their credit, electric cars are quiet and pollution-free. That is the reason why they do have an assured future. Some governments have already legislated for zero-emissions vehicles to be compulsory: both California and Switzerland insist that a certain percentage of cars sold from the year 2000 have to produce no toxic emissions whatsoever. By 1990, there were over 500 electric and solar road cars registered in Switzerland. But there is still some debate as to whether the electric car is really more 'green' or simply shifts the pollution from the vehicle to the power stations from which it derives its energy.

As a result of the weight of lead-acid batteries, most electric cars have had to be small to retain a practical performance and range. There have been exceptions, but from the early days, the electric car has always been an important part of the microcar world.

Electric power, virtually abandoned since the turn of the century, came back to popularity during the Second World War, as there was precious little petrol available. Particularly in France, a whole host of small electric vehicles popped up, only to disappear again immediately following the war.

Once again, electric propulsion entered a dark age. There were some odd refugees from the golf courses which could be registered for road use in California, USA during the 1950s and 1960s but it was not until the 1970s and renewed fears of an oil crisis that electric power was seriously considered again.

Several firms, most in the USA, produced electric vehicles during the 1970s — and with some success. In Holland, there was even an attempt to introduce a hire system for electric cars in Amsterdam. But as soon as the fuel supplies came back and looked like staying, suddenly everyone realised that electric cars still suffered from the same impracticalities as ever.

It was not until the late 1980s, when global warming and the influence that cars were having on the greenhouse effect was fully appreciated, that electric came back on a tide of green issues. The formula remained the same: usually small cars powered by lead-acid (or now silver) batteries, promising much but not really delivering. They were just as expensive as before, but at least there was some public groundswell of opinion which actually put its money where its conscience gave vent.

Big manufacturers offered electric versions of their small cars: Fiat had an electric Panda and Peugeot an electric 205. Others offered prototypes, not as publicity 'concept cars', but as serious propositions for production: BMW showed the E1 and General Motors had the sporting Impact, both scheduled for full production.

There were many small firms beginning to offer electric cars, too. Particularly in Switzerland, the range of electric cars on offer is becoming bewildering and there are even circuit races for electric cars. The Swiss have also pioneered the marketing of solar-powered cars for the road and there is now an annual Electric and Solar Car Show incorporated into the Geneva Motor Show.

But the question remains: will there be the long-awaited breakthrough in battery technology to make electric power a true replacement for oil? Until that happens, legislation notwithstanding, the internal combustion engine will continue to be the main power source for cars.

PEUGEOT VLV

Wartime France: there was no fuel available for private use and so the market was ripe for any alternative. Electric power provided the answer for many drivers. Peugeot essayed the VLV in June 1941, a simple four-wheeled open car powered via a 1.3 bhp SAFI electric motor. Its total weight was 800 lbs (365 kg), of which 350 lbs (160 kg) consisted of the batteries. Its top speed was a modest 22 mph (36 km/h), but it had a useful range of 50 miles (80 km). Production continued until February 1945.

Peugeot's wartime VLV was typical of the myriad electric cars produced in occupied France.

SCAMP

The Scamp was one of many electric car projects of the late 1960s which was intended for eventual production but never made it. The first prototype was completed by Scottish Aviation of Ayrshire, Scotland, in 1965 but the car's debut was not until 1967.

Scottish Aviation's Scamp was a serious proposal for an electric production car in 1967.

A simple glassfibre-bodied two-seater just under 7 feet long, it used twin series-wound ventilated DC electric motors which, it was claimed, could take the car to a top speed of 35 mph. Its maximum range was 26 miles.

Ten of the twelve Scamps built were allocated for evaluation by various Electricity Boards, the remaining two being kept by Scottish Aviation. Despite vague plans to productionise the Scamp as "the commuter car of the 'seventies", to sell at around £350, nothing more came of this short-lived project.

FORD COMUTA

The Ford Comuta of 1967 was claimed to be the "first to be designed and developed by a major motor manufacturer specifically for electric propulsion." It was a product of Ford's Research and Engineering Centre, in Dunton, Essex, and it was exceptionally short. Disappointingly, it could travel 40 miles only at a top speed of 25 mph and Ford relegated it to the 'experimental out-takes' file.

The Comuta was Ford's first attempt at an electric city car. Two could easily fit within the length of a Ford Cortina.

FORD BERLINER

Ford tried its hand again with an electric prototype in 1970 with the Berliner, this time produced by Ford of Germany. It was surprisingly pretty and could seat two adults in front and two children in the rear, who gained access via a tailgate. At only 7 feet (213 cm) long, it was very compact but again was intended only as an experiment.

The rear passengers in the Ford Berliner prototype entered through a rear door and sat facing each other.

VOITURE ELECTRONIQUE

La Voiture Electronique is a simple title; but then this was a simple car, perhaps the simplest ever built. It consisted of a plastic triangle with moulded-in seats, between which sat a single joystick that controlled acceleration, braking and steering. Hence the car could be driven from either seat — but there was a fool-proof system which stopped the car dead if the driver attempted to do the impossible with his joystick...

Simplicity itself: La Voiture Electronique was stripped down to the bare essentials of transportation.

Twin Jarrett motors drove the rear wheels. The car weighed only 440 lbs (200 kg), and could reach just 15 mph (25 km/h) over a distance of up to 40 miles (60 km).

Built in Moselle, France, from 1968, the simplicity of the concept was diluted as the years elapsed, so that the 1972 'leisure' version, oddly called the Porquerolles, had such luxuries as bumpers, indicators and wing mirrors. The Cab version even had a hard-top! The Jarrett brothers who conceived the car were unable to keep it in production beyond 1976.

Later versions of the Voiture Electronique were more kitted out for the road: this is the 1972 Porquerolles.

CROMPTON-LEYLAND

Leyland's Crompton-Parkinson Division was the major British producer of electric vans in the early 1970s. It did once attempt a passenger car with a pretty little Michelotti-styled two-seater, shown at the 1972 Geneva Motor Show. Powered by no fewer than 24 batteries, it could reach 33 mph (53 km/h) and had a range of 40 miles (65 km). But it weighed one third more than a Mini (at 1,983 lbs) and was not the practical production proposition its designers had hoped for.

Michelotti styled this Crompton-Leyland, an electric prototype from Leyland's truck division.

SEBRING VANGUARD

This was probably the most persistently marketed electric car ever made. Its origins lie with a golf cart made by Club Car Inc of Augusta, Georgia. Vanguard Vehicles of Kingston, New York, developed a two-seater road version in 1972 which was frankly pig-ugly. But at $2,000, it did at least sell in reasonable numbers (10 a week in 1973).

Following a move to Sebring, Florida, the firm now sold a new model with the name Sebring Vanguard. This was much prettier and had a closed GRP body on a steel chassis. Six 6-volt batteries gave it a top speed of 28 mph (45 km/h) and a range of 50 miles (80 km). Quite why the Vanguard should have become so popular is not clear, but with a production rate of 20 cars per day, it was probably the best-selling electric car ever. Over 2,000 were sold between 1974 and 1976 alone.

The firm's name changed to Commuter Vehicles in the late 1970s, but the product remained the same. It did acquire 'federal' bumpers which looked ridiculous on a car so short (they made it 10 feet – or 305 cm – long) and there were taxi and van versions with a lengthened wheelbase. There was an option of solar panels for the van, which would increase range by 10% after eight hours' charging.

But by the beginning of the 1980s, the Commuter was beginning to look like the spark had gone out of it. Although many thousands were sold, the company did not last beyond about 1986.

America's Sebring Vanguard was one of the most popular electric cars ever: this is the long-wheelbase Comuta with its outsized bumpers.

ENFIELD

No other British electric car came as close as the Enfield to series production. But in the end, it hardly scratched the commercial surface.

The story began with the Electricity Council which, in 1968, asked firms to tender a design which met its criteria for an electric car. The successful tenderer was a Greek shipping magnate called J K Goulandris, who built his electrically-powered two-seater in 1969, and dubbed it the Enfield 365. His intention was to manufacture certain parts in Britain and build the actual vehicles on the Greek island of Syros.

In the event, he was persuaded to build his Enfield entirely in Britain. After many tests by the Electricity Council, an order was placed for 66 cars to a slightly improved specification, the new car being called the Enfield 8000. The first cars were delivered in 1975.

The 8000 was 8 inches shorter than a Mini (284 cm), but seated only two passengers. Even using lightweight aluminium body panels, the all-in weight was massive: just short of a ton (975 kg). With eight 12-volt batteries humming away, the Enfield could reach a top speed of 40 mph (64 kmh) and drive up to 56 miles (90 km) on a single charge.

As well as supplying the Electricity Council, the Isle-of-Wight-based factory took the bold decision to offer the car for public consumption. The Enfield's big problem was its cost: an outrageous £2,808 in 1975 — this was at a time when the Mini cost £1,298 and you could buy the basic Ford Granada for considerably less. The development of an 8000-based four-seater Moke-style car called the Runabout did not help sales. Only around 40 cars were sold to the public.

The first prototype from Enfield, the 365, was built in 1969. It looked promising.

Part of the fleet of Enfield 8000s used by the Electricity Council in Britain.

Enfield also essayed the Runabout, a compact electric Mini Moke-style four-seater based on the 8000 chassis.

The Enfield passed the crash test with flying colours.

LEPOIX

French-born engineer Louis Lepoix was dabbling with microcars as early as 1950, when he built a coupé version of the German-built Champion. He was also an electric car enthusiast and, in 1974, he created the Urbanix, a squarish electric prototype which was not destined for production.

However, his next projects were. The Lepoix Shopi and Ding were debutants at the 1975 Frankfurt Motor Show (Lepoix was based in Baden Baden, Germany). Despite appearances, the Shopi was a four-wheeler, but the front pair sat virtually side-by-side on what must be the shortest front axle ever seen. The rear wheels were powered by a 1.5 kW 24-volt motor and the batteries sat under the two seats. As its name suggests, the Shopi was intended as an errands car and had a top speed of just 6 mph. It was also an extremely compact vehicle: only 59 inches (150 cm) long.

The same glassfibre body was used in the Ding, which eschewed the Shopi's steel subframe for an extraordinary external chassis which looped up and around the whole body. Together with the three bubble-form wheels, the impression was of a refugee from a science-fiction film set. It was also more racy, with a top speed of almost 16 mph (25 km/h)!

Alas, the odd couple were never to reach intended production in 1977. The author can only express his regret that the public was not ready for such a prodigy as the Lepoix Ding.

Louis Lepoix's Shopi was one of the world's smallest ever cars, intended as a local runabout.

If the Shopi was for shopping, Lepoix's Ding was for... dinging? The madcap Ding was more than a show folly: it really was scheduled for production.

ELECTRACTION

ElecTraction of Maldon, Essex, was the only serious electric car project from Britain other than the Enfield. Run by ex-Ford and British Leyland man Roy Haynes, the firm's first project was the EVR-1 (Electric Vehicle Research), which it displayed at the 1976 Earls Court Motor Show. There it claimed (perhaps optimistically) that it could travel at 40 mph (64 km/h) for up to 100 miles (160 km). It also said that, with battery technology progressing so fast(!), "within ten years, we expect EVR to travel 500 miles at speeds up to 85 mph"...

The EVR-1 did not enter production, but it did form the basis of two follow-up projects. The most obviously EVR-inspired was the Tropicana, basically a drop-top version of the EVR-1 and first seen in 1977. Twelve 6-volt batteries accounted for its heavy weight of 2,156 lbs (980 kg) and it was claimed to do 35 mph. It made use of an MGB windscreen and Citroën 2CV headlamps.

The other model was the Rickshaw, again a vehicle intended for leisure use, but this time sporting a 4/5 seater glassfibre body. Again it used a 7.5 hp electric motor and a dozen 6-volt batteries and the weight remained at about one ton (1,020 kg), although the Rickshaw was longer at 10 feet 11 inches (332 cm). All ElecTractions used Vauxhall Viva steering, suspension, wheels and axles.

A full range of weather equipment was offered. The version with a frilly Surrey top was known as the Bermuda and the doorless version was called the Campus. A full hard top was also available, as was a pick-up/van which was called the Trusty.

AC Cars, which had just been forced to halt production of the Invacar invalid carriage, was all lined up to produce the ElecTractions in series, at a rate of no less than 2,500 cars in the first year, mostly intended for sale in the USA. Production never began, largely due to the expense of putting the cars through Type Approval and, despite all good intentions, ElecTraction ceased its activities in 1978, having built fewer than twenty vehicles.

ElecTraction's EVR-1 prototype of 1976: note the quirky but functioning running boards and mock-Edwardian rear lights.

The Rickshaw produced by ElecTraction was intended for use in leisure resorts, universities and so on.

ZAGATO

Virtually concurrent with its radically rebodied Ferraris and the styling of Bristol's 412, the celebrated Milanese coachbuilder, Zagato (founded 1919) was working on a tiny electric city car, about the most extreme polar opposite ever seen in the motoring world.

The Zele was more than just a design exercise for Zagato: it launched them into their only large-scale car manufacturing project of the 1970s. (Zagato had previously made special bodies on other manufacturers' chassis).

Only 77 inches (195 cm) long, the Zele was a tall glassfibre-bodied two-seater of quite pretty appearance. Front suspension was independent but at the rear there was a solid axle, on to which was mounted a Marelli 1 kW electric motor, powered by four 24-volt batteries. Weighing just over half a ton (550 kg), the Zele 1000 (standing for the wattage) was capable of 25 mph (40 km/h) and could run up to 43 miles (70 km). The Zele 2000 could reach 38 mph (60 km/h).

The Zele was the production version of an earlier design study, the Milanina. Although the Zele was first shown at the Geneva Motor Show of 1972, it did not reach production until 1974. By this time, the American electric car specialist, Elcar, had shown sufficient interest for Zagato to agree to export the Zele to the 'States. An unlikely British importer was found in the luxury car maker Bristol. Production was running at a healthy 225 examples for 1975.

A long-wheelbase van version arrived in 1976 and, by popular demand from the USA, a doorless golfing version was developed. In 1978, Zagato showed a prototype for its replacement Zele, with squarer, more modern, lines. By the time it reached production in 1981, it was much bigger and so was not a suitable replacement for the Zele: it was called the Nuova Zele and sold alongside the old model.

As the 1980s progressed, Zagato began to concentrate again on its work as a *carrozzeria*, with Aston Martin, Nissan, Alfa Romeo and Lancia. The Zele models took a back seat and were finally withdrawn from sale in about 1991. Several thousand had been made.

Coachbuilders Zagato took an unexpected direction with the Zele, a tiny electric two-seater produced from 1974.

The Amsterdam traffic problem was eased in the 1970s by the hire-it-as-you-go Witkar fleet.

WITKAR

Following on from the Provo group's anti-pollution campaign in the 1960s in Amsterdam, when it released hundreds of white bicycles into the city for people to use freely, there was an attempt to do a similar project with electric cars. The so-called Witkars (white cars) were telephone-kiosk shaped electrically-powered two-seaters. They were run by a central agency and subscribers would have access to keys and recharging stations. The project worked for a few years, but most people still preferred to use their own cars.

PILCAR/CARVILLE

(photo on p.175)

No other country has such stringent concern for vehicle emissions as Switzerland. As a result, it has a full catalogue of electric cars. One of the first was the Pilcar, a pretty city car instigated by Victor Perremond in collaboration with l'Electricite Neuchateloise and the Societe Romande d'Electricite. The car was actually designed and built by Swiss coach-builder Franco Sbarro.

It arrived in 1977, offering good performance (up to 56 mph or 90 km/h) from a 22 bhp motor in a good-looking package about the size of a Mini — but it suffered from being overly heavy at almost 1¾ tons (1,700 kg). It could seat four people in its three-door polyester body. Alternatively, a pick-up was available.

Priced at 16,000 francs, it was a rare example of a commercially available electric car, selling 10 examples in its first year. The name was changed to Carville in 1979 and the company which marketed it was called Vessa. But it was too early to catch the green tide and the last year it was available was 1983.

MOBILEK

The Mobilek was the brainchild of Dudley-based inventor Doreen Kennedy-Way. She hoped to lauch her electric trike in 1979, intending it for shopping trips and commuting by businessmen. It was to cost £1,000, but there were problems with the Ministry of Transport's type classification for the car and it never went into production.

SINCLAIR C5

Sir Clive Sinclair should surely have known better? His infamous Sinclair C5, which he financed from his own pocket, was an unmitigated disaster and all but sunk him as an entrepreneur.

To its credit, the C5 cost only £399 and could be driven legally by 14 year-olds. It was also a truly environment-friendly vehicle, its battery-and-pedal power offering a range of up to 20 miles. The trouble was, it was not exactly driver-friendly. The hapless C5 owner had to sit exposed to the elements, to exhaust fumes and, worst of all, to advancing juggernauts, with his hands fumbling under his buttocks attempting to steer it. And he had to pedal up hills, as the battery couldn't cope!

Sinclair aimed to produce no less than 100,000 C5s from a Hoover factory within the first year. But things got off to a bad start when it snowed heavily on launch day, in January 1985, and the C5 turned into a toboggan. By February, several local authorities had banned its use amid fears about driver safety.

In April, production halted due to 'a gearbox fault', but it quickly became obvious that disastrously poor sales were the real reason for the stoppage. Sinclair put the project up for sale in June, found no buyer, and so left the receiver to pick up the pieces in October. Only a few thousand had been supplied.

The Mobilek trike of 1979 was optimistically intended for commuting to and from work as well as shopping trips.

The infamous Sinclair C5 even had its own magazine: in reality, drivers felt extremely vulnerable on the road.

PINGUIN

Switzerland leads the world in productionised electric cars and the Swiss/ Hungarian Pinguin Euromobil became the first ever solar-powered car to be registered for road use in Germany (in 1988). The two-seater could reach a top speed of 31 mph (50 km/h) and, presuming the sun did not shine, would carry on running for up to 62 miles (100 km).

This Pinguin electric car became the first solar-powered car to be officially registered in Germany in 1988.

MINI-EL

Denmark has made a significant mark in electric car production with the 1987 Mini-El, which succeeded where that other Danish project, the Hope Whisper, had failed. In appearance, it was rather like a grown-up Sinclair C5 but in practice it was an awful lot better. Its electric propulsion was superior and its bodywork — available in 'Hatchtop' and 'Cabriolet' versions — was far more substantial.

The firm received a boost when a batch of Mini-Els were ordered for the Barcelona Olympic Games in 1992. In Britain, where the models were imported, the name had to be changed to City-El after Rover objected about the use of the name Mini.

The Danish Mini-El was one of the first of a 'new generation' of electric cars which are actually finding buyers.

RENAULT ZOOM

Along with Peugeot (which productionised an electric version of its 205) and Citroën (which produced the Citela project), Renault also became active in electric car development in the early 1990s. Its contribution was the whacky Zoom concept car of 1992.

Developed jointly with Matra, the Renault Zoom's most striking feature was the way it tucked its rear wheels under its short body to ease parking. In a car measuring only 8 feet 10 inches (270 cm) at full stretch, this made the Zoom extremely short — almost two feet (60 cm) shorter.

The high-tech Zoom featured an electric engine which allowed a 95 mile range and a top speed of 75 mph. It was claimed to be 90% recyclable. The two passengers gained entry via gullwing doors.

Production was never even a possibility, although Renault stated, probably not very seriously, that the Zoom aimed to "provide pointers to the future of the small city car."

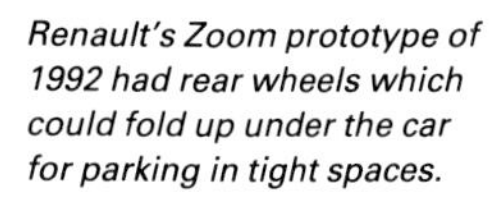

Renault's Zoom prototype of 1992 had rear wheels which could fold up under the car for parking in tight spaces.

CHAPTER SEVEN

ODDBALLS & MIGHT-HAVE-BEENS

TO many eyes, all microcars are oddballs. In truth, it is not an easy task to find many cars *more* oddball than a bubble car. Luckily, they do exist. Not just a few of them, either, but a whole profusion so completely whacky and obscure that it could hardly fail to enthrall the lover of trivia. There just seem to be more truly miscellaneous microcars than any other type of car.

So with no more ado, here is the 'dustbin chapter', chock full of all the indefinable out-takes from the world of microcars, the lost causes, the bizarre stunts, the mystifying prototypes: in short, the short and short of it!

BIZARRE PROTOTYPES

City cars never really stood much chance in Italy after the Fiat 500 was launched. The only examples to have scored any success were either very small or very bizarre.

A rotating cockpit was Marquis Bargagli's solution for parking problems with his Urbanina.

Into the latter category came the Urbanina, probably the only ever example of a swivelling car. Designed by the Marquis Piergirolamo Bargagli, it was first shown at the 1965 Turin Motor Show. It was composed of two sections: a lower unit (into which was fitted the engine and running gear) and an upper body which could rotate at will to allow the driver to exit at any point around the car. Some examples even had wickerwork bodies.

Only 6 feet (183 cm) long, it could be powered by either a Bosch electric engine or an Innocenti 198 cc air-cooled two-stroke 3.75 bhp engine. Its total weight was 558 lbs (254 kg). Ambitious production plans were announced and the Urbanina appeared at many shows over the following years, at least until 1973, latterly with a conventional body, but nothing much ever came of the project.

One of the smallest cars ever made was Stuart Smith's Gecko of 1966. It measured all of 5 feet 6 inches (169 cm) long and weighed just 250 lbs (114 kg). Its 200 cc engine was capable of powering the car to a heady top speed of 50 mph (80 km/h). Despite hopes to put the car into production, the Gecko remained a one-off.

Bob Collier's microcar prototypes never made it into production (although the intention was always there). Of the five or so cars he built during the 1960s and 1970s, many had interesting features. One had infinitely variable belt gearing and the only controls on board were the steering wheel and 'stop' and 'go' pedals. Braking came on automatically as the vehicle slowed. Its engine was a 273 cc Briggs & Stratton industrial unit. Another used a 350 cc Kohler engine which could be swivelled through 360 degress along with the single front wheel.

Probably the world's smallest four-wheeled car: the 1966 Gecko remained a prototype.

This is the Colliday designed by Bob Collier and piloted, by the looks of it, by the singers from the B-52s.

Whacky Sibona & Bassano Tsetse featured a forward-hinging canopy and an empty space where the engine should have been!

One of the obscurer Italian coachbuilders of the 1960s was Sibona and Bassano. Their quite outlandish Tsetse was an obscurity *par excellence*. Shown at the 1964 Turin Motor Show, it was a plastic-bodied two-seater with a flip-forward canopy. Sibona and Bassano rather hopefully suggested that any engine of about 250 cc could be fitted in their engineless prototype. Tragically, no-one took them up on it.

Quite why the British firm Universal Power Drives, manufacturers of the lovely Unipower GT sports car, ever got involved with as whacky a project as the Quasar-Unipower is an absolute mystery.

In 1968, Vietnamese-born fashion designer Quasar Khanh designed a bold new vehicle which he persuaded UPD to build. Yet the element of design was conspicuous mainly by its absence, for where is the stylist's hand in the creation of a transparent mobile cube?

A strong tubular steel chassis contained widened Mini subframes, with a Mini Clubman engine sited beneath the rear passengers' bottoms. Glassfibre panels covered the chassis, while a substantial steel tube frame rose up to hold huge sheets of toughened glass in place. Entry was via sliding patio doors at either side (or on some versions through the front) and the four passengers sat on transparent plastic seats, ensuring that absolutely everything was on show. A green-tinted roof struggled to prevent the interior becoming a greenhouse.

Shorter than it was wide, narrower than it was high (5 feet 4 inches by 5 feet 6 inches by 6 feet 2 inches), the little cube was a dreadful car to drive unless you were obsessed, to the exclusion of all else, by good visibility. It was flat out at 50 mph; not that you'd want to go at that speed in a vehicle which promised shattering shards of glass in your face at the slightest incident.

Six Quasar-Unipowers were built by UPD during 1968. What became of these exhibitionists' delights is transparently unclear.

A transparent cube designed by a Vietnamese-born French fashion designer and built in Britain: the Quasar-Unipower.

PROTOTYPES FROM LARGE CAR MAKERS

Daihatsu Fellow Buggy. 1968 was the year of the beach buggy. Daihatsu couldn't resist doing this to its Fellow microcar for the 1968 Tokyo Show. The result was a most unlikely fun car.

General Motors. The American car giant built five microcar prototypes in the late 1960s, four of which are pictured here: they were powered by various engine combinations: the 511 (left) had a 1100cc petrol engine, the 512 (third left) was electric and the others were petrol-electric and petrol-driven.

Zagato Zanzara. This pretty open two-seater was initially presented as the Zagato Hondina with a Honda N360 engine sitting in the tail. A couple of years later, it reappeared as the Zanzara transplanted with Fiat 500 running gear. The entire front of the car, including the windscreen, tilted forward.

Daihatsu BCX-II/Toyota Town Spider. A Japanese trait for creating implausible show cars was typified in this co-project from Toyota and Daihatsu. Shown at the 1973 Tokyo Show, this electrically propelled glass house was about as likely to go into production as an all-in-one tie-and-underpants combination.

Ford Ghia City Car. This prototype measured only 8 feet 6 inches (250 cm) long and used a 1-litre engine mounted in the rear. The four-seater was never intended as a production car.

Suzuki CV1. If ever there was an attempt by a major manufacturer to revive the bubble car, the Suzuki CV1 of 1981 was it. A single-seater, it was powered by a 50 cc engine which, in a car weighing only 330 lbs (150 kg), gave it a fuel consumption of 140 mpg. Amazingly, Suzuki built a batch of 50 'sample' cars.

Ghia Trio. This was "Ford's smallest ever four-wheel passenger car." Shown at the 1983 Geneva Show, it was only 7 feet 9 inches (236 cm) long and weighed 743 lbs (338 kg). Its 250 cc engine returned 70 mpg. The three passengers sat in an 'arrowhead' formation.

Volkswagen Scooter. At Geneva in 1986, VW displayed the three-wheeled gullwing Scooter. It measured 10 feet 5 inches (318 cm) long but had distinctly un-microcar-like performance from its 90 bhp VW Polo-derived engine: 136 mph (218 km/h) top speed and 0-60mph in under 8.5 seconds.

A LOST CAUSE

The famous British motorcycle maker BSA might have returned as a car manufacturer if this prototype had been given the go-ahead. It was built in 1960 with a Triumph Tigress 250 cc vertical twin engine. The extremely compact open two-seater body was hand-made from a single sheet of steel. However, it was not approved for production and the Ladybird was the last car BSA ever built.

The cute BSA Ladybird prototype might have relaunched BSA's car manufacturing side.

SOME JAPES

This was a picture which appeared in *Small Car* magazine in 1963. The basic lines of the Scootacar can be made out, but why is there a Rolls-Royce grille attached? In fact, this was a spoof article parodying the uncritical testers' styles of the weekly magazines *Motor* and *Autocar*. The fictional vehicle was titled the 'Rolls-Canardly Silver Spoon', which was the imaginary car BMC created when it "took over both Rolls-Royce and Scootacar." Incidentally, it got a rave review.

Rolls-Canardly Silver Spoon!

STARLETS OF THE SILVER SCREEN
A Heinkel was a co-star of the British comedy classic *I'm Alright Jack*, based around the adventures of the (surely over-developed?) St Trinian's School pupils. At the wheel is George 'Arthur Daley' Cole. Would you buy a second-hand bubble car from this man?

The Heinkel in the film I'm Alright Jack.

Wahay! In the British TV series, *The Chinese Detective*, a Bond Bug was involved in the unlikeliest car chase ever, when it careered after a Messerschmitt KR200. The plot surrounded a microcar collector and many of the cars used came from the Register of Unusual Microcars. Both the Bug and the Messerschmitt were destroyed during filming!

Messerschmitts have enjoyed numerous TV and film appearances. One of the oddest was Terry Gilliam's *Brazil*, in which it starred as a 'personal transport pod' dressed up with extraneous appendages. It ended up getting burnt out.

FROM THE BAROQUE TO THE SPACE-AGE: SOME AMERICAN ODDITIES

The Rollsmobile was the first of many American replicas of Edwardian carriages which emerged in the late 1950s and early 1960s and truly began the 'replicar' age.

The idea came from Smarts Manufacturing Co in Fort Lauderdale, Florida, who acquired the blueprints of the 1901 Oldsmobile, then in the hands of General Motors, and concocted a ¾-scale replica of the car in 1958. It certainly looked the part with its mahogany-overlay body and 20 inch chrome-plated wheels, but it came equipped with nothing more than a 4 bhp industrial engine and an automatic gearbox and was capable of about 30 mph (48 km/h) and 100 mpg.

It could be driven on highways but most were bought by hotels and garages. Unlike most veteran replicas, the Rollsmobile stayed in production until well into the 1970s. It was even improved upon: the firm later brought out a 1902 version!

The 1958 Rollsmobile: a baroque reproduction of a pioneer of the motoring age in miniature.

The outlandish SP Spi-Tri was a seriously lengthened Bond Bug bodyshell fitted with a choice of electric or petrol engines.

Why anyone would do this to a Bond Bug is a mystery. Its lines can just be made out in the oversized SP Spi-Tri from Structural Plastics of Tulsa,

Oklahoma. The name hints at the fact that it was a GRP monocoque car, not a separate chassis design as the original Bond had been. Early versions had electric power, weighed a ton and cost $12,000. Later versions could be ordered with petrol engines, but it is doubtful whether very many people bought one.

Designing what amounts to a shoebox with wheels stuck on all four corners may not sound very clever, but, styling apart, there was a lot more to the Monocoque Box than that.

Of course it was a product of Californian brains. The Box was designed by Dan Hanebrink and Matt Van Leeunen of Costa Mesa as a 'multi-purpose vehicle'. To that end, it was four-wheel-drive, four-wheel steering *and* amphibious.

The body was a compact balsa wood/glassfibre sandwich, the lower half of which was the 'chassis', although by the time the upper body was bonded on, the car lived up to its name as a 'Monocoque'. Hydropneumatic suspension and A-arms which swang and pivoted on all four corners were high-tech features of the 1977 car.

The prototype used a 500 cc rear-mounted Kawasaki two-stroke engine which was capable of propelling the Box to a top speed of 100 mph. Later versions used VW Beetle and Honda Accord engines. There was just a single disc brake on the left front-drive axle. The Apex gearbox was reputedly able to make the Box travel as fast in reverse as in forward!

The Box measured 129 inches (328 cm) long and only 44 ins (112 cm) tall. Entry to the two-seater cockpit was via the upward-hinging windscreen. Boxes were offered for sale as kits from $2,210 or more if you ordered the optional body panels required to transform it into an amphibian. Implausible? Undoubtedly, but all true, I promise!

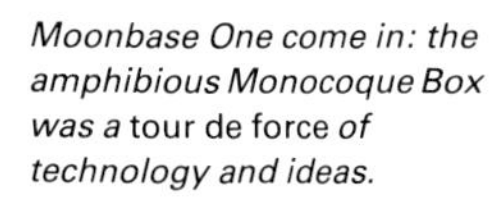
Moonbase One come in: the amphibious Monocoque Box was a tour de force *of technology and ideas.*

WHAT IS THE WORLD'S SMALLEST CAR?

In a book about microcars, the question inevitably arises, what *is* the smallest car ever made? There is no one answer: it depends on the parameters you set. But here are some of the most compact cars ever seen.

Ignoring such flights of fancy as motorised skateboards and even motorised skates (both of which have been marketed) the smallest vehicle offering seats was almost certainly the Peel P50 made on the Isle of Man (see Chapter 2). Peel stated that it measured 50 inches (127 cm) long by 33 ins (84 cm) wide by 46 inches (117 cm) high and weighed a mere 132 lbs (60 kg).

By contrast, the German Brutsch Mopetta (Chapter 2) measured 67 inches (170 cm) long by 34.6 inches (88 cm) wide by 39 inches (100 cm) high and its weight has been quoted as being between 134 lbs (61 kg) and 172 lbs (78 kg). A later electric car, the Lepoix Shopi (see Chapter 6), was only 59 inches (150 cm) long by 43 inches (110 cm) wide.

Battle of the tiddlers: Peel P50 just edges out the Brütsch Mopetta (left) as the smallest car ever made.

There have been narrower cars: the Thaon AT2 measured only 80 cm wide, while the Larmar was just 72.5 cm wide. This even beats the 'wheels in' parking width of the Reyonnah (75 cm).

Many microcars have used moped engines of less than 50 cc capacity. The early Fend Flitzer used a 38 cc Victoria motorbike engine. But the production car with world's smallest engine was probably the Crystal Engineering Sachs Trice, which had a 30 cc unit. One home-built special, the Lynch three-wheeler, used a 22 cc industrial engine.

FIFTY THINGS YOU NEVER KNEW ABOUT MICROCARS

1 While testing his **Flitzer** in the Alps, Fritz Fend lost his brakes and was forced to stop his runaway car with shoe leather.

2 If the door latch on an **Isetta** is not closed properly, and the front door should ping open *en route*, the steering wheel hinges up with the door...

3 In the English brochure for the **Kleinschnittger**, it is described as having "rubber band suspension."

4 The **Staunau K400** of 1950 was the most underpowered car in history: it had a 389 cc engine pulling a 13 feet 6 inch (412 cm) long body.

5 A **Messerschmitt** was driven from Southampton to John o' Groats and back in the 1950s. It covered the 1,505 miles at an average speed of 33 mph.

6 Ten years before Alec Issigonis coined the word Mini, Lawrie **Bond** was building his Minicar.

7 A **BMW-Isetta** was converted for use on railway lines as a Rail Taxi.

8 Many microcars such as the **Messerschmitt** had reversible Dynastart on their engines, allowing for travel just as fast in reverse as forwards.

9 Microcars claimed to be amphibious include the **Trippel SK9**, the Argentinian **Leeds** and the **Brütsch Mopetta**.

10 The only known rotary-powered microcars were the **NSU Sport Prinz** and the **Rocaboy Kirchner**.

11 The **Champion** had front and rear wings pressed from the same pattern.

12 There have been two microcars with five wheels: the 1919 **Briggs & Stratton Flyer** and the French **CEDRE 5x5**.

13 **Porsche** made some microcar prototypes in 1955: one had a tiny 597 cc three-cylinder four-stroke engine.

14 In 1951, the Italian coachbuilder **Boneschi** built a body in the shape of a toothpaste tube on a Fiat 500 Topolino chassis.

15 There was 'another' **Citroën**: Andre's cousin Joseph built his own electric prototype called the Velocar.

16 There have been numerous examples of micro plagiarism: for instance, the **Hoffman** was a close copy of the Isetta and the **Julien MM7** mimicked the Rovin D3.

17 **Mahag** built a miniature 125 cc replica of the VW Beetle with a fold-up canopy like the Messerschmitt.

18 **NSU** made several bubble car prototypes from 1952 to 1954, with engines from 250 cc to 750 cc.

19 The most-licensed car ever produced was the **Fuldamobil**, made in at least seven countries outside Germany.

20 There have been several attempts to make an enclosed motorbike retaining only two wheels, such as the 1952 **Rhiando** and 1976 **Quasar**, both British.

21 The 1950 **Schwammberger** had a single steering rear wheel to which the engine was also attached.

22 Ex-*Dr Who* John Pertwee used to drive a **Messerschmitt**.

23 The longest-lived microcar design is the **Tempo**, originally introduce d in Germany in 1936 and reputedly still in production in India.

24 The Czech-built **Prvenac** not only had four wheels placed in a diamond formation, its chassis was also articulated.

25 **Volvo** built a 790 cc prototype in 1954 which never came to production.

26 The American **Scoot-mobile** was built from spare Air Force bomber fuel tanks.

27 The list of British firms which made large-engined sports cars at the same time as making microcars includes **AC, Allard, Bond, Fairthorpe, Reliant** and **Lea-Francis** (the latter being involved with **Nobel**).

28 The 1948 **Alamagny** from France had four wheels in a diamond pattern and twin headlamps stacked vertically, one on top of the other.

29 During 1964, an **Isetta** was used to smuggle nine people (one at a time!) across the Berlin wall by squeezing them into the heater compartment. The ruse was only discovered when a passenger sneezed at the checkpoint!

30 Queen Elizabeth II is reputed to have ridden in a **Messerschmitt** at the Mountbatten Estate.

31 There was an all-glassfibre 4 cwt van version of the **Isetta** built in Britain.

32 In Spain, there were apparently instances of **Biscuters** getting stuck head-to-head in steep mountain gulleys as no reverse gear was fitted and neither driver could push his car back up the hill.

33 Paul **Kleinschnittger** built a moped with an air mattress which made it amphibious.

34 A converted **Reliant Regal** van was used in the Amazon for ecology research in 1977.

35 The 1956 prototype of the Hillman Imp (called the **Slug**) used a tiny flat-twin engine in the tail.

36 **Trabants** were circuit-raced in Hungary.

37 A 'car' was made out of the cockpit of a **Blenheim** aircraft and actually driven on the road between 1947 and 1954.

38 One **Collier** prototype had a single steering front wheel but disguised the fact with an additional two dummy front wheels on outriggers, making it look like a four-wheeler.

39 The world's best-selling sub-700 cc cars were the **Trabant** (3,690,099 sold) and the **Fiat 500** (well over 3 million made).

40 Microcars have been made in Guernsey, Isle of Man, Isle of Wight, San Marino and Liechtenstein.

41 In 1965, Jim Parkinson circumnavigated the globe in a pre-war 98 cc **Rytecraft Scootacar**.

42 **Dennis Adams** designed a prototype of a city car which could stand on its end for parking.

43 At the launch of the **Bond Bug** in 1970, one journalist rolled a 'Bug' completely over!

44 The American microcar firm **Zoe** built its own version of the Reliant Rialto.

45 Motorbike maker **Yamaha**'s first-ever car was the 50 cc moped-engined PTX-1 prototype of 1983.

46 A number of large car makers, notably Volkswagen and BMW, are preparing new 'minicars' for launch before the next century.

47 The **DD** had the unlikeliest production career, starting life in Vietnam, then moving to Morocco.

48 There were 'beach car' versions of numerous micros, including the **Fiat 500**, **600**, **Multipla**, **850**, **BMW 600** and **Vespa**.

49 The **Meyra 55** of 1950 actually had its engine sitting exposed *inside* the single-seater cockpit.

50 The tiny American **Eshelman Sportabout** of 1953 was mysteriously offered with an optional snow-plough attachment.

Glas Goggomobil (left).

Bugatti Type 56 electric car.

A-Z OF MICROCARS

THE following includes all microcars produced, or intended for production, from 1940 to the present date. Included are all cars of up to 700 cc engine capacity, plus cars with larger engines if they are exceptionally small (under 280 cm long). Not included are commercial vehicles, two-wheeled vehicles, racing cars and invalid carriages.

Make & Model	Engine: cc/bhp	Date of manufacture	Length (cm)	Country of origin
Abarth 600	633-747/-47	1955-60	321	I
500 Gran Turismo	479/20	c1957-63	297	I
500 GT Zagato	479/23	1959	—	I
595/595SS	594/27-32	1963-71	297	I
695 SS	689/38	1964-71	297	I
AC Petite Mk I	346/7.75	1953-55	312	GB
Petite Mk II	353/8.25	1955-58	312	GB
ACAM Nica	125-218/—	1984-87	227	I
Galassia/Zeta	50/—	1987	—	I
Acoma Mini-Comtesse 770	50/2	1976-80	174	F
Super Comtesse	50/3	1978-84	180	F
Star/Starlette	50-124/3	1982-84	228	F
Comtesse Sport/Coupé	50/3	1982-84	215	F
[**Addax**: see Vitrex]				
[**Ad-Hoc**: see Arie]				
ADI	120/4.2	1950	—	D
[**AEMS**: see Inter]				
Aero Minor	615/13-20	1946-53	404	CS
Aerocarene	684/23	1947	343	F
[**Aeromecanique**: see Voisin]				
AFA	Electric	1943	—	E
AG Traction Cabrio	602/31-50	1987-date	—	P
[**Aichi**: see Cony]				

Airway	—/10	1949-50	—	US
AISA Pullman-Biscuter	197/—	1955	283	E
Aixam 50-325D/400D	50-400/—	1984-87	260	F
325i	276-325/—	1987-92	257	F
400i	400/10	1987-90	298	F
600i	600/13	1988-92	270	F
500i	470/12	1990-92	298	F
400i Twin	400/5	1991-92	298	F
A540/Record	276/5	1992-date	255	F
A550 Twin	400-479/5-15	1992-date	302	F
Alamagny	569/13	1948	342	F
Alba Regia	350/—	1955	—	H
Albrecht	175/9	1950-51	256	D
Aleu Bambi	200/—	1954	—	E
Allard Clipper	346/8	1953-c55	307	GB
All Cars Charly/Snuggy/ Tobrouk	49-250/—	1974-85	210	I
Allemano 600	633/22	1955	—	I
Alta	198/10	1968-77	315	GR
AM	250/9	1948	280	CS
American Buckboard	—/25	1955-56	—	US
API Rickshaw	150-325/—	1955-date	—	IND
APP TXA/Sprint	325-505/—	1990-91	250	F
Ardex T-53	100-125/4	1953	—	F
Typ M	50/2.5	1954-55	220	F
Arie Hopi	125/—	1984-85	236	F
Arola 10/11/12/SP	Elect/47-125/3	1976-82	184	F
18	50/3	1982-83	200	F
Minoto	49-125/—	1982-83	230	F
Arzens L'Oeuf	Electric	1942	—	F
Carrosse	125/—	1951	—	F
Astra	322/15	1956-60	290	GB
[**Atlas**: see Livry]				
[**Atomo**: see Samca]				
Attica	198/10	1964-68	315	GR
Audax	692/20	1947-48	—	CH
Autelec	Electric	1942-43	—	F
Autobianchi Bianchina	479-499/13-21	1957-70	302	I
Fiat 500 Giardiniera	499/18	1968-77	318	I
Autocykl	500/12	—/—		CS
Autoette	Electric	—	—	USA
[**Autofioretti**: see Fioretti]				
Automirage Pick Wick	594/23	1975-c83	—	I
Mirage 3	50-250/2.5	1976-85	212	I
Mirage S3R/S4R	125-250/—	1982-85	227	I
[**Autonacional**: see Biscuter]				

Auto-Riksha	—/—	1980s	—	IND
[**Auto Sandal**: see Otosantaru]				
[**Autozodiaco**: see All Cars]				
Avia	350/15	1956-57	—	CS
Avolette	125-250/5-19	1955-58	240	F
AWS Shopper/Piccolo	247/14	1971-76	289	D
[**Badal**: see Sunrise]				
[**BAG**: see Spatz]				
Bajaj Tempo Autorickshaw	454/10	1960-date	—	IND
Autorickshaw	145/5.5	1980s	—	IND
Balaton	250/—	1956	—	H
Balbo B400	398/14.9	1953	—	I
Baldi Frog	125-595/—	1973-75	220	I
Bambi	—/—	1960s	310	RCH
[**Bambino**: see Hostaco]				
Bamby	49/—	1983-84	—	GB
Banner Boy Buckboard	—/2.75	1958	—	USA
Basse	250/—	1953	—	D
Basson's Star	—/—	1956	—	USA
Belcar	200-250/—	1955-c56	330	CH
Bellier Formule 85	50-325/3	1980-87	235	F
XLD/GTD/VX	325/500—	1987-date	250	F
VX650	650/—	1994-date	250	F
Bel-Motors Veloto C-10	49/1.85	1976-c81	—	F
Veloto C-10S	49/1.85	1979-c81	—	F
Super Veloto C12/Minoto	49/1.85	1980-c81	220	F
Benelli BBC	600/21	1949-52	—	I
Bergemann	200/5	1953	335	D
Berkeley Sports 322	322/15	1956-57	310	GB
Sports 328/B65	328/18	1957-58	310	GB
B90	492/30	1958-59	310	GB
B95/B105	692/40-50	1959-60	324	GB
T60	328/18	1959-60	310	GB
Bertoni	124/6	1948-49	—	I
Biscuter	197/9	1953-60	257	E
Biscuter 2000	50-360/—	1984-c85	—	E
Bjelka Squirrel	500/15-30	1956	—	SU
Blohm & Voss	—/—	1945-46	—	D
BMA Hazelcar	Electric	1952-57	280	GB
BMA Amica	50-223/1.5-12	1971-80	213	I
Brio	47/3	1978-c86	186	I
Nuova Amica	50-360/4-11	1980-date	232	I
BMW 331	490/13	1949-50	—	D
Isetta 250	247/12	1955-57	228	D
Isetta 300	295/13	1956-62	228	D

600	585/19.5	1957-59	290	D
700	700/30-40	1960-65	387	D
BMW	Electric	1949-1970s	—	US
Bobbi-Kar	—/7-10	1945	—	US
Bohler	147/4	1946-47	200	D
Boitel Minicar	400-589/12-18	1946-49	315	F
Bond Minicar Mk A	122-197/5-8	1949-51	274	GB
Mk B/C/D	197/8	1951-57	297	GB
Mk E	197/8	1957-58	335	GB
Mk F/G	246/12	1958-66	335	GB
875	875/34	1965-70	328	GB
Bug	700-750/29-31	1970-74	279	GB
Boneschi 500 C	569/16.5	1951	—	I
600 Spyder	633/22	1955	—	I
[**Bonnallack**: see Minnow]				
Borgmann Dwerg	—/—	1950s	—	NL
Boselli Libulella	160/6	1952	—	I
Bouffort City-Car/Enville	500/—	1951-60	215	F
BRA CX3	500-650/50-64	1992-date	—	GB
Breguet	Electric	1942	—	F
Brissonet	200/—	1953	—	F
Brodeau	Electric	1942-43	—	F
Brogan	—/10	1946-51	—	US
Brütsch T	125/9	1950-51	250	D
400	400/10	1952	312	D
200 Spatz	191/10	1954	330	D
Zwerg	191-247/10-14	1955-57	240	D
Zwerg Einsitzer	74/3	1955-56	220	D
Mopetta	49/2.3	1956-58	170	D
Rollera	98/5.2	1956-58	210	D
Bussard	191/10	1956-58	330	D
Pfeil	386/13	1956-58	330	D
V2	98-247/5.2-13.5	1957-58	255	D
V2N	479/15	1958	315	D
BSA Ladybird	250/—	1960	—	GB
Bubu Cabin Scooter	50/—	1982	—	J
Buckaroo	—/—	1957	—	US
Buckle Dart	295-394/15-18.5	1959-61	289	AUS
Budai	—/—	1948	—	H
Bugatti Type 68	350/40	1945-48	—	F
Bunger	600/19	1947-49	—	DK
Burgers	400/13	1951	312	NL
[**Burgfalle**: see Spatz]				
[**Butenuth**: see Econom]				
[**B&Z**: see Electra King]				

Californian Commuter	90-200/10.5	1985	302	US
Calstart SEV	Electric	1992-date	—	US
Canta	633/22	1955	—	I
Caprera Algol	Electric	1969	—	I
Carter Coaster	Electric	c1966	—	GB
Casalini Sulky/David	50-325/3.3	c1975-c87	188	I
Bretta	50/3.3	1980-82	237	I
CC Zero	652/30	1991-94	—	GB
CCA Condesa/Duquesa	602/29	1982-c86	—	E
CEDRE Mini 1/Midinette	Electric	1974-87	185	F
Centuri TXA/Sprint	325-505/—	1992-93	250	F
CGE	Electric	1941-46	—	F
Chadwick 300	289/13	1960s	221	US
Champion ZF	196/5	1946	198	D
Ch-1	196/5	1949	242	D
Ch-2	248/6.5	1949-50	260	D
Ch-250	248/6.5-10	1950-51	285	D
Ch-400	396-398/14-15	1951-54	318	D
500 G	452/19	1954-55	340	D
Chapeaux	Electric	1940-41	—	F
Charbonneaux	425/18	1955	—	F
[**Charlatte**: see Hrubon]				
[**Charly**: see All Cars]				
Chatenet La Chatelaine	325-654/—	1985-date	262	F
Stella 211	411/5	1994-date	250	F
Chausson CHS	330/9	1948	—	F
Cheetah	Electric	1992-date	254	CH
Cicostar	50-123/3.2-5	1980-83	220	F
CIMEM Girino	—/4.5	1951	—	I
Motospyder	125/—	1955	—	I
Cingolani	125/3	1952	260	I

Calstart SEV.

Citeria	582/19.5	1958	—	NL
Citroën 2CV	375-425/9-18	1948-68	378	F
2CV4	435/24	1968-80	383	F
2CV6	602/22-31	1963-90	383	F
Dyane/Ami	435-602/24-32	1961-82	387	F
Bijou	425/12	1959-62	394	GB
Mehari	602/29	1968-87	352	F
LN/LNA	602-652/32-36	1976-85	343	F
Visa Special	652/36	1978-85	369	F
Citroën Monocar/Velocar	70/—	1946/1954	250	F
[**City-El**: see Mini-El]				
[**City Mobil**: see Jephcott]				
[**Claeys**: see Flandria]				
Clauzet	425/18	1959	—	F
Clearfield	569/16.5	1950s	—	US
Cleco	Electric	1940	—	GB
Clua	247-350/—	1955	—	E
	497/17	1958-59	—	E
CMV	Electric	1944-46	—	E
Colenta	Electric	1986	—	D
Colliday Commuter/Chariot	50-350/—	1960s	—	GB
Colt Economy	380/—	1958	—	US
Comet	—/4.5	1946-48	290	US
Comet	—/6	1950-51	—	US
[**Commuter**: see Sebring Vanguard]				
[**Comtesse**: see Acoma]				
Condor S-70	677/32	1957-58	360	D
Convenient Machines Cub	—/—	1983	—	US
Cony	200-354/18.6	1952-67	299	J
Cooper	500/—	1947-52	—	GB
Corat Lupetta	250/5.5	1946	260	I
Coronet	328/18	1957-60	365	GB
[**Crayford**: see Willam]				
Crofton	722/35	1959-61	—	US
Crosley	725-742/26.5	1939-52	368	US
Crystal Sachs Trice	30/—	1992-date	—	GB
Cubster	—/6.6	1949	—	US
Cursor	49/—	1985-87	284	GB
Cushman 82	—/—	1982-83	—	F
CVC Bedouin	602/29	1985-c87	—	GB
Dacia 500 Latsun	500/—	1987	300	RO
DAF Daffodil	590/19-22	1957-61	361	NL
Dagsa	500/—	1951-52	—	E
Dagonet 2CV	425/18	1952-57	—	F
Daihatsu Bee	540/13.5	1955-?	—	J

Midget/Tri-Mobile	250/12	1950s	288	J
Hi-Jet	356/17	1960-66	299	J
Fellow	356/23	1966-70	299	J
Fellow Max	356/31-37	1970-76	299	J
Cuore/Mira/Leeza	547-659/28-63	1976-date	320	J
Opti	659/55	1992-date	330	J
Dalat Jeep/Berline	602/29	1970-c80	—	VN
Dalnik	615/19.5	1948	—	CS
Danger Autino	125/—	1947-49	—	D
Libelle	250/6.3	1949/1953	—	D
Danilo	175/—	1958	—	DK
Daulon	250-350/—	1950	—	F
Daus	—/—	1954	—	D
David	175-345/—	1950-57	280	E
Daytona	—/2	1956	—	US
DB	610/30	1952	—	F
DD	125/—	1949-50	—	VN/MA
De Carlo 600	589/19.5	c1960	290	RA
Decolon	125-175/5-10	1957	309	F
DECSA Lisa	50-250/—	1982-c87	240	RSM
Delsaux Modulo	47/3	1980-83	184	F
[**Delta**: see Champion]				
Deltamobil	197-250/9.5	1954-55	—	D
De Pontac	425-500/—	1955-57	—	F
Deshais	125-350/6-15	1950-52	310	F
Diavolino	125-750/—	1984-c91	210	CH
Diehlmobile	—/3	1962-64	—	US
DIM	594/23	1977-83	320	GR
Dinarg D-200	191/10.6	1959-69	243	RA
DKR	—/45	1954	—	DK
DKW F7/F8	584-692/18-20	1939-53	—	D
F89P	692/23	1950-54	—	D
Doddsmobile	—/—	1947	—	CDN

The DECSA Lisa was the pride of the people of the tiny principality of San Marino, where it was made in small numbers.

Dolo	571/23	1947-48	365	F
Dornier Delta	197/9.5	1955	288	D
Minimo	50/—	c1957	—	D
Delta II	392-643/20-23	1969	210	D
Delta II G	493/20	1971	220	D
Delta 2	Elec 782/27.5	1972-73	220	D
D-Rad-Rikscha	496/12	1951	280	A
Drolette	Electric	1969	—	CH
DS Malterre	125-175/6-8	1955	—	F
Duport Caddy/510D	510/10	1977-83	224	F
511/611/Parco	510-602/10	1980-88	292	F
311/312	359/6	1983-85	244	F
Parco 50 Diesel	359/6	1986-88	244	F
360	359/6	1988-91	244	F
400 A	400-510/10	1988-89	244	F
125/4	654/13	1987-92	310	F
D50 GL/GTV	327/6	1990-92	248	F
Onyx	385/6	1990-92	252	F
Eagle	Electric	1970s	—	US
Eaglet	Electric	1948	—	GB
[**ECAM**: see Vitrex]				
Econom Teddy	245/6.5	1950	215	D
Edith	198/1.5	1952-56	259	AUS
EEC Worker's Playtime	250/10	1952-54	267	GB
Egan	—/—	1952	—	GB
Elcar Cinderella/Zagato	Electric	1970s	195	US
ElecTraction EVR	Electric	1976	—	GB
Rickshaw Bermuda	Electric	1977-78	332	GB
Tropicana	Electric	1977-78	—	GB
Electra King	Electric	1961-c80	256	US
Electric Shopper	Electric	1964-c73	—	US
Electrociclo	Electric	1945-46	—	E
Electrolette	Electric	1941-43	—	F
Electro Master	Electric	1962-c73	—	US
Electro Motion	Electric	1975-76	350	US
Electro-Renard	Electric	1943-46	—	F
El-Jet	Electric	c1992-date	—	DK
Elpo	125/4	1948	—	CS
Enfield 365	Electric	1969	—	GB
8000	Electric	1975-76	269	GB
Runabout	Electric	1976	—	GB
Entreposto Sado 550	547/28	1982-86	236	P
Erad Capucine	47/3	1975-80	199	F
Capucine/Duo	123-290/6-7.5	1980-83	197	F
Capucine/Plein Air	49-400/6	1983-88	232	F

EEC Worker's Playtime.

Midget	123-600/7.5	1982-date	270	F
6.50D/6.125D/6.500	290-505/—	1986-date	249	F
Junior 300/400	Elec-304-400/—	1988-date	224	F
6.400D	400/—	1989-date	272	F
Spacia	Elec/304-505/—	1991-date	257	F
Agora	Elec/505/5-12	1992-date	210	F
Erla-Bond	125/—	1950-52	274	DK
Eshelman Sportabout	—/8.4	1953-60	—	US
ESO T-250	250/—	1959	—	CS
Esoro E 301/H301	Electric	1993-date	307	CH
Espenlaub	396/—	1952	350	D
[**Eufradif**: see Marden]				
FA Kabrio-Roller	175/9	—	—	D
FAF	602/29	1978-c83	359	RI/CI/SN/P/RCA/PG
Fairthorpe Atom	250-650/11-35	1954-57	328	GB
Atomota/Atom Major	650/35	1957-60	328	GB
Falcon	602/31	1984-date	—	GB
FAM	500/21	1952	—	I
Farmobil	700/35	1960-?	—	GR
FART	499/17.5	1965	—	I
Faure	Electric	1941-47	—	F
Favel	Electric	1941-44	—	F
Felber	350-398/12-15	1952-54	275	A
Fend Flitzer	38-98/2.5-4.5	1948-51	200	D
FK150	148/6.5	1953	282	D
Feora	175/22	1981	—	US
Fermi Lucertola	368/14	1948	—	I
Ferves	499/22	1967-c73	263	I
FH	197-324/9-15	1956-60	—	E
FIAM Johny Panther	124/10	1978-c80	239	I
Fiat Topolino 500A	569/13	1936-42/1946-48	322	I
500B/C	569/16.5	1948-55	324	I
600/600 Multipla	633/22	1955-60	322	I
500	479/13.5	1957-60	297	I
500D/F/L/R	499-594/18-21	1960-75	297	I
500 Giardiniera	499/17.5-21.5	1960-77	318	I
126	594-652/23-30	1973-87	305	I
126 Bis	704/25	1987-date	311	PL
Panda 30	652/30	1980-86	338	I
Fiat Poker	652/24	1982-84	—	GR
Fiberfab Scarab STM	450/—	c1975-c79	—	US
Sherpa	602/29	1975-c80	—	D
FIMER Lucertola	250-358/7-14	1947-50	—	I

Fioretti F50	50/—	1978-c81	234	I
Fissore 600 Gran Luce	633/22	1955	—	I
Flandria	250-400/—	1953-54	—	B
Flipper I	47/3	1978-80	200	F
II/Donky	50/2.6	1980-84	213	F
III	50/3.1	1982-84	244	F
[**FMR**: see Messerschmitt]				
Frada	97/3	1948	—	CS
[**Fram King**: see Fuldamobil]				
Frankl Autoroller	250/—	1949	—	D
Fredcar	602/31	1987-c88	—	F
Frisky Sport	249-324/15-18	1957-61	286	GB
Family Three	197-250/9.5-11	1959-64	278	GB
Prince	324-328/15	1960-64	—	GB
Sprint	492/30	1958	311	GB
FSM Polski-Fiat 126	594-704/23-30	1975-date	305	PL
106 Beskid	—/—	1985	—	PL
Fuji Cabin	123/5.5	1957-58	—	J
Fuldamobil	248/8.5	1951-52	272	D
N-2	359/9.5	1952-55	285	D
S-1 (NWF 200)	197/9.5	1954-55	297	D
S-2	359/9.5	1954-55	297	D
S-3	191/10	1956	297	D
S-4/S-5	191/10	1955-56	310	D
S-6	191/10	1956-57	310	D
S-7	191-198/10	1957-69	315	D
Gabry	150-250/6-13	1963	—	I
GAD 500	496/23	1953	—	PL
Gaitan	—/—	1950s	—	E
Galassi	250/6.5	1949	—	I
Gallati	—/—	1983	—	US
Galy	175-280/11-18	1954-57	295	F
Gashopper	49/—	1980s	224	US
Gaslight	—/4	1960-61	—	US
Gateau Egzo-7/Egzo-3/Maximini/				
Break/Grande	Elec-49-500/-	1983-88	230	F
Vison	325/6	1988-91	245	F
Forum	325/6	1991-92	245	F
GE	Electric	1979	234	US
Gecko	200/—	1965	169	GB

Friskysprint

Galy (right)

Giannini 750 Sport	569-747/16-28	1953	—	I
500 TV	499-570/21-39	1964-75	297	I
500 4-cyl	698/—	1965	297	I
590 GT	586-598/31-41	1964-75	297	I
650 NP	652/35-40	1970-75	297	I
500 R GT	694/34.5	c1973-75	297	I
Sirio	652/35	1972	—	I
350 Economy Run	390/16	c1973-75	297	I
126 GP	594-794/29-41	1973-83	305	I
Panda	652/36	1981-82	341	I
Gilcolt	700/31	1972	—	GB
Gill Getabout	322/15	1958-61	290	GB
[**Girino: see CIMEM**]				
Glas Goggomobil	197-247/14	1954	270	D
Goggomobil T250	247/14	1955-56	290	D
Goggomobil T300/400	296-392/15-20	1955-69	290	D
Goggomobil TS300/400	296-392/15-20	1957-69	304	D
T600/T700/Isar	584-688/20-30	1958-64	343	D
GMT Rivelaine	47-49/—	1981-83	242	F
Gnom	123-143/5.5-7	1950	220	D
Goliath GP700	688/25	1950-55	415	D
GOM Chihuahua	50-125/—	1974-75	—	I
Gommelel	147/3	1947	—	D
Gordon	197/8	1954-58	350	GB
Grewe & Schutte	197/9.5	1954-56	—	D
Grignani Parva	125/5	1950-53	—	I
[**Grunhut**: see Belcar]				
Gurgel Itaipu	Electric	1976-80	260	BR
280M/BR-800/Supermini	792/33	1988-date	280	BR
Gutbrod Superior 600	593/20-22	1950-54	356	D
Superior 700	663/26-30	1951-54	356	D
Superior Sport	663/30	1951-52	360	D
Haargaard	250/15	1950s	—	DK
Hanomag Partner	697/28	1951	400	D
Hansan	489/15	1958	—	NL
[**Hans Vahaar**: see Fuldamobil]				
Harborough	Electric	c1967	—	GB
Hartnett	594/19	1951-57	—	AUS
Heathfield Slingshot	500-650/50-64	1993-date	—	GB
Heinkel 150	174/9.3	1956-58	255	D
153/154	198-204/10	1956-58	255	D
Heinkel	198/10	1957-59	255	RA
Heinkel-I	204/10	1958-60	255	IRL
Helicak	—/—	1970s	—	RI
H-M Free-Way	Elec 348-453/-	1979-c83	292	US

Hoffmann Kabine	248/12.5	1954-55	228	D
Honda S500/S600	531-606/44-57	1962-66	330	J
N360/N400	354-401/31-33	1966-74	300	J
N500/N600	500-598/40-45	1968-74	300	J
Z/Z600	354-598/31-38	1970-74	300	J
Life	356/30	1971-74	300	J
Vamos	354/30	1971-74	300	J
Today	545-656/31-52	1985-date	330	J
Beat	656/64	1991-92	330	J
Hope Whisper	Electric	1983-c87	362	DK
Hoppenstand	350/8.5	1947-49	—	US
Horlacher City	Electric	1993-date	220	CH
Hostaco Bambino	197/9.5	1952-c57	—	NL
Hotzenblitz	Electric	1992-date	270	D
Hrubon	50-998/3-39	c1978-89	208	F
[**Humbee Surrey**: see Mitsui]				
Huracan	197/—	1958	—	E
Hurst 100	97/2	1946	250	D
250	248/6	1948-50	300	D
Husqvarna	500/—	1943	—	S
HWS	400/18	1955	—	D
Hybricon Centaur	Electric	1979-c81	—	US
Ibana	547/28	1994	242	SF
IFA F-8	692/20	1948-55	—	DDR
[**IFG**: see Gnom]				
IMP	—/7.5	1949-55	—	US
IMP	645/40	1960-61	—	I
Impala	499-652/20-30	1981-85	—	GB
Impuls 77	Electric	1977	—	S
Incamp	—/—	1952	—	E
Indesten	594/23	1974-75	—	I
Innocenti 500/650/Small	548-659/31	1982-93	316	I
Inter	175/8.5	1953-56	295	F
Internationale	Electric	1942	—	NL
Iota	350/29	1951	—	GB
Irat	500/—	1950/1953	—	F
ISO Isetta	236/9.5	1953-55	228	I
ISSI Microbo	125/6	1952-54	220	I
Jamos	643/25	1964	—	A
Jarc Little Horse	250/13	1955-56	—	GB
Jarrett	Electric	1969-71	168	F
Jawa	350/—	1953-56	—	CS
JB Minor	—/5	1949	—	AUS
JBF Boxer	602/31	1992-date	—	GB

JDM 49 SL/125	49-325/—	1981-84	—	F
Parthenon	50-325/—	1984-85	239	F
Nueva	325/—	1985-88	—	F
SLD	49-325/—	1985-86	213	F
Furio/	325-654/—	1988-88	290	F
X5	317-502/—	1988-date	247	F
Orane	Elec/317-502/5	1994-date	257	F
Jehle MX Safari	602/29	1979	—	FL
Jephcott Micro	—/—	c1983	259	GB
Jet	197/—	1955	—	E
Jipy Cabriolino	480/16	1989-date	—	E
JLC Jorgia	602/22	1972-73	—	E
[**Johnny Wigo**: see FIAM]				
Johnsonmobile	—/3	1959	—	US
Jolly	—/—	1984	—	I
Joseso	200/10	1959	—	RA
JPB	602/29	1985-date	—	P
Julien MM5	310/8	1946-50	285	F
MM7	325-368/10	1949-53	299	F
Junior	197/8	1955-56	—	E
JZR	500-650/50-64	1989-date	299	GB
Kapi Kapiscooter/Chi-Qui	124-175/4	1950-55	—	E
Topolino	197/8	1952-58	—	E
Kattewitz	—/14	1958	—	PL
Keinath	—/—	1949	—	D
Kelly	—/—	1945-46	—	GB
Kendall	594/25	1945-46	360	GB
Kersting	123-197/1.5-10	1950	250	D
[**Kewet**: see El-Jet]				
Kieft	650/35	1952	—	GB
Kikos	49-125/—	1980-83	—	F
King Midget	380-480/8.5-12	1946-69	—	US
Kleine Wolf	197/3	1950	280	D
Kleinschnittger 98	97/3	1949	220	D
F125	123/4.5-5.5	1950-57	290	D
F250	246/15	1954-57	305	D
Kleinstwagen	348-398/14	1952	—	B
[**Kover**: see Livry]				
Kover Capilla	200/6	1951-52	—	E
Krajan	250/—	1948	—	CS
Kreibich	250-350/—	1948-49	—	CS
Kroboth Allwetterroller	174-197/9-10	1954-55	280	D
Kurier	250/9	1948	—	CS
Kurogane	—/18	1957-62	—	J
KVS Mini/Gad'Jet	49-123/3-9	1976-c84	210	F

Lamb-Kar	250/9	1950	—	GB
Lambretta FDC 150 Rickshaw	150/—	1957-?	—	I
[**Lambretta**: see Willam]				
Lambro	125/—	1952	—	I
La Novi	Electric	—	—	F
Larmar	249-350/8-10	1946-51	228	GB
Lawil Varzina/City	125-246/12	1967-86	207	I
Log	246/15	1973-c76	207	I
Berlina A4	246/15	1984	238	I
Le Dauphin	Elec 100-175	1941-42	—	F
Leeds Microcoupe/Autoneta	324/—	1960-64	—	RA
L'Electra	Electric	—	—	F
Le Nivet	500-680/—	1947	—	F
Leonard La Marysa	325-500/—	c1993-date	—	F
[**Le Piaf**: see Livry]				
Lepoix Shopi	Electric	1975-77	150	D
Ding	Electric	1975-77	255	D
Leterer	375/9	1955	—	B
[**Libelle**: see Danger]				
Libelle	199/8.5	1952-55	—	A
Lidowka	498/14	1945-50	300	CS
Lightburn Zeta	324/16	1964-67	307	AUS
Zeta Sports	493/20	1964-67	323	AUS
Ligier JS 4/6/8	50-125/3.2	1980-85	197	F
330	327/—	1985-86	197	F
Serie 5	327/—	1986-87	197	F
Serie 7	325-654/—	1987-91	249	F
Optima	50-265/4	1989-date	250	F
JS 16	505/25	1994-date	250	F
Livry Kover	150-175/—	1949-51	245	F
Atlas	150-200/—	1949-52	277	F
Le Piaf	175/—	1951	—	F
Lloyd LP300	293/10	1950-53	310	D
LP400	386/13	1953-57	345	D
LP250	250/—	1956-57	310	D
LP600/LS600/LT600	596/19-24	1955-62	335	D
Alexander/Alexander TS	596/25-30	1958-62	335	D
Alexander Frua	596/25	1959	400	D
Lloyd	650/17.5	1948-51	373	GB
Logo	Electric	1993-date	—	CH
Lomar Honey	125/—	1985-86	360	I
Lomax	602/31	1983-date	345	GB
Lucciola	250/8	1948-49	300	I
[**Lucertola**: see FERMI]				
[**Mada**: see Danger]				

Mahag	125/4	1950	255	D
Maico MC 400	398/14	1955-56	320	D
MC 500G	452/18	1955-56	340	D
MC 500	452/18	1956-57	343	D
500 Sport	452/20	1957	375	D
Maliutka	—/—	1965	—	SU
[**Mallalieu**: see Microdot]				
Manocar	125/4	1952-53	225	F
[**Marathon**: see SIOP]				
Marden Espace/49	Elec/49-435/3-24	1975-83	220	F
Fetta/Mini/Maxi	49-400/5	1983-87	230	F
Channel	325-400/5	1987-91	246	F
Alize	325-400/5	1988-92	250	F
Marketour	Electric	1964-c73	—	US
Marland Jorgia	602/29	1971-76	—	F
Marold	300-350/—	1951	—	A
Martin Stationette	—/—	1954	—	US
Mathis VL333	707/15	1946	340	F
Mazda R-360	356/16	1960-70	298	J
Carol	358/20	1962-72	299	J
P600	586/28	1964-c66	—	J
Chantez	359/33-35	1972-c78	299	J
Carol	547-657/38-61	1989-date	325	J
AZ-1	657/64	1992-date	330	J
MCA	594/26	1983-date	351	GB
McKee Sundancer	Electric	1972-c80	305	US

McKee Sundancer.

NSU Prinz.

[**MDP**: see Vitrex]				
Meduza	300/—	1957	300	PL
Meister G5N	49/3.5	1969	230	A
Melex Electrocar	Electric	c1971-79	—	PL
Mendizabal	125/—	1952	—	F
Merry '01	—/4	1958-62	—	US
Messerschmitt KR175	174/9	1953-55	282	D
KR200/201	191/10	1955-64	282	D
Tg500	493/20	1957-64	305	D

Meyra 48/50	197/3	1948-51	—	D
53	197/7	1953-55	—	D
55	197/7	1950-52	252	D
55 Zweisitzer	248/8.5	1952-53	285	D
200	197/9.5	1953-55	350	D
200-2	197-350/10-14	1955-56	345	D
[**Microbo**: see ISSI]				
Microcar RJ/DX	50-124/—	1980-84	235	F
50/125/300/600	50-600/—	1984-88	248	F
Solea	50-125/—	1983-85	238	F
Spid!	273-654/13	1988-91	258	F
Lyra	Elec/273-505/—	1990-date	258	F
Newstreet	50/5	1994-date	258	F
Microdot	Elec-547/28	1976-81	—	GB
Mikasa	585/18-21	1957-61	381	J
Mikrus	300/15	1956-60	301	PL
Minicar	308/7.5	1948	—	CS
Mini-Cat	49/2.2	1978-80	158	F
[**Mini-Comtesse**: see Acoma]				
Mini-El	Electric	1987-date	—	DK
Minima	602/32	1970s	240	F
Mini-T	Electric	1992-date	240	US
Mink	198/—	1968	—	GB
Minnow	200-250/10	1951-52	—	GB
Mipal	250/9	—	—	CS
Mirda	250/9	—	—	CS
Mitsubishi A10	493/21	1960-63	314	J
Minica/Skipper	359/18-38	1962-72	299	J
Colt 600	594/25	1962-65	—	J
Minica/Ami	359-546/30-38	1972-84	299	J
Minica/Econo	546/31-42	1984-93	320	J
Minica/Toppo	657-659/40-64	1993-date	330	J
Mitsui Humbee Surrey	285/11.5	1950-62	—	J
MiVal Mivalino	172/9	1954-56	282	I
MM4	—/—	1945-46	—	F
Mobil	350/—	1954-57	—	CS

Murrill.

Mobilek	Electric	1979	—	GB
Mobilette	Electric	1965-c73	—	US
Mochet CM	50-125/—	1947	240	F
Velocar	100/3	1948-51	—	F
CM 125/175	125-175/3.5-7	1951-58	264	F
Monocoque Box	500/—	1977-79	328	US
Monterosa Toi et Moi	633/22	1955	—	I
Moraly	250/16	1956	250	F
Moravan	350/—	1956	—	CS
Moretti La Cita	350/14-41	1945-50	300	I
600	592/18-20	1950-57	340	I
500D Coupé	592/20	1952-54	—	I
500 Special Bodies	499/21-30	1959-75	—	I
Minimaxi	499-594/20-23	1970-c81	—	I
Morin Aerocar	125/—	1948	—	F
II	400/—	1957	—	F
Motobecane	125/—	1974	—	F
Motoflitz	500/12	1949	380	D
Movilutil	—/15	1959	—	E
MT	125-175/7	1955-57	—	E
Muller-Niedhart	—/—	1952	—	S
Munguia	300-400/15-18.5	1957-67	—	E
Murrill	Electric	1981	274	US
MV	—/—	1952	—	I
MVM	325/18	1956	—	GBG
MW Minimobil/Cityboy	50/—	1984-?	—	D
MYMSA	175-500/8.8	1957	—	E
NAMCO Pony	602/29	c1975-86	366	GR
Nardi Nardyna	610/35	1952	—	I
[**Neckar**: see NSU-Fiat]				
[**New Map**: see Rolux]				
NJ	—/12	1954-56	—	J
Nobel 200	191/10.2	1959-62	315	GB
Nobletta	191/10.2	1960-62	320	GB
Noval	50-231/—	1982-83	225	F
NSU Prinz I	583/20-30	1958-60	315	D
Prinz II	583/20-30	1959-60	315	D
Prinz III	583/23-30	1960-62	315	D
Prinz 4	598/30	1961-73	351	D
Sportprinz	583-598/30	1959-67	357	D
Wankel Spider	497/50	1964-67	357	D
NSU-Fiat Jagst	633/20-23	1956-69	322	D
Weinsberg	499/15-22	1959-63	310	D
[**NWF**: see Fuldamobil]				

Ohmic	356/11.5	1950s	278	J
Oltcit Axel	652/34	1981-84	372	RO
Opperman Unicar	225-328/18	1956-59	295	GB
Stirling	424-493/16-25	1958-59	—	GB
Orix	610/27	1952-54	—	E
OTI Microcar	125/—	1957-59	287	F
Otosantaru Auto-Sandal	350/5.5	1947-54	240	J
[**P70**: see Zwickau]				
P80	250/12	1957-58	245	PL
Panda	582/—	1955-56	—	US
Panhard Dyna	610/18-28	1946-52	382	F
Panther	480-520/12-14	1954-57	356	I
Pappenburger	600/22	1953	—	D
Paris-Rhone	Electric	1947-50	—	F
[**Parva**: see Grignani]				
Pashley	197/6	1953	—	GB
Pelican Rickshaw	600/—	1953-57	—	GB
Pasquini Valentine	251/12	1976-c80	310	I
Passat	200-600/22	1952-53	360	D
PB	175/—	1955	—	F
Peel Manxman	250-350/—	1955-56	229	GBM
P50	49/4.2	1962-66	135	GBM
Trident	49-100/4.2	1965-66	183	GBM
[**Pente**: see Weiss]				
Perl Champion 250	250/9	1951-54	280	A
SK 10	610/28	1953-54	360	A
Peugeot VLV	Electric	1941-45	267	F
PGE	Electric	1979-c84	271	I
[**PGO**: see Puli]				
Phoenix	197-250/—	1955-56	—	ET
Philipsons	692/15	1946-48	—	S
Piaggio Scooter	280/20	1990	—	I
Pichon-Parat Vespa 400	400/14	1959	285	F
Piccolino	650/12-14	1987-date	210	CH
Pilcar/Carville: see Vessa]				
Pinguin	197/9.5	1953-55	342	D
[**Pinguin** 4 Euromobil: see Puli]				
PIO	396/12	1949-50	300	D
Pioneer	Electric	1959	—	US
Pionier	496/23	1953	380	PL
[**Piotrowsky**: see PIO]				
Poinard	125-175/4	1952	—	F
Poirier	98-125/—	1928-58	—	F
Powerdrive	322/15	1955-58	—	GB
Praga	—/9	1952	—	CS

Precicar	49/—	1983	—	F
Precico Funcar	340/24	1975	—	CDN
[**Projet Plus**: see Silaos]				
Prvenac	250/—	1959	—	YU
PTV	250/10	1956-62	282	E
Publi-Retro	602/31	1984-85	—	F
Publix	—/1.75-10.5	1947-48	183	US/CDN
Puck	600/22	1952	390	D
Pulcino	124/5	1948-50	260	I
Pulga	—/—	1952	240	E
Puli	Electric/270/—	1987-date	246	H/CH
Puma Minipuma	760/30	1975	266	BR
Puma	49/4	1980-82	216	F
Pup	—/7.5-10	1947	260	US
P. Vallee Sicraf	175/5	1952-57	—	F
Chantecler	125-175/5-8	1956-57	310	F
Pypper	50/—	1986	246	E
Quasar-Unipower	1098/48	1968	163	GB
Quincy-Lynn Urba Car	Electric/—	1975-c85	—	US
Urba Electric	Electric	1977-c85	320	US
Urba Trike	Electric	1978-c85	—	US
Trimuter	Electric —/16	1980-c85	366	US
Rago	325/14	1967	—	U
Ramses	583-598/20-30	1959-73	—	ET
Rapid	200-350/8	1946-47	304	CH
Rappold 500 C	569/16.5	1954	—	I
Ravasi	700/21	1947	—	I
REAC	610/26	1953-54	—	F/MA
Reliant Regal MkI-VI	747/16	1951-62	312	GB
Regal 3/25	598/25	1962-67	343	GB
Regal 3/30	701/30	1967-72	343	GB
Rebel	598-748/25-31	1964-73	351	GB
Resine	—/—	1954	—	I
Revelli	Electric	1941	—	I
Reyonnah	175-250/11	1950-54	290	F
[**Riboud**: see Vitrex]				
Rieju	—/—	1953	—	E
Riesenacker	600/—	1949	—	D
Ringspeed Eco 2	957/45	1989-92	267	GB
[**Rivelaine**: see GMT]				
RNW	197/8	1951	264	GB
ROA	197/—	late 1950s	—	E
Rocaboy Kirchner	Electric	1972-c75	285	F
Rodley	750/20	1954-56	278	GB

Rollera	125/6	1959	212	F
Rollsmobile	—/3	1958-c73	—	US
Rolt LM49	49/—	1982-83	200	F
Rolux Baby VB 60	125/5	1946-49	265	F
VB 61	175/6	1949-52	—	F
Romanazzi Mototaxi	—/—	1953	—	I
Romi-Isetta	236-300/9.5	1956-61	228	BR
Rosenbauer	250/10	1950	320	A
[**Roulante**: see Brissonet]				
Roussey	700/—	1948-50	—	F
Rovin D2	425/10	1947-48	280	F
D3	425/11	1948-50	306	F
3 CV D4	425-462/10-13	1950-61	315	F
Rox Mini-Jeep Roxy	122-190/—	1986-87	240	F
[**Ruhr**: see Pinguin]				
Ruspacromo	Electric	1984-?	—	I
Russon	197-250/—	1951-52	326	GB
Ryca	434/12	1956	280	RA
Rytecraft Scootacar	98-250	1934-40	—	GB
Saipac Baby-Brousse/Jyane	602/29	1968-c90	—	IR
Sam	—/—	1954-56	—	PL
Samca Atomo	246/10	1947-51	307	I
SATAM	Electric	1941	—	F
Saviano Scat	—/25	1960	317	US
Savio Albarella	499/20	1967	—	I
Jungla	767/32	1970s	—	I
Jungla	594/23	1974-c83	302	I
[**Sbarro** Carville: see Vessa]				
Scamp	Electric	1967	211	GB
[**Schmitt**: see Hrubon]				
Schwammberger	—/—	1950	—	D
Scoiattolo	499-594/20-23	1969-c82	297	I
Scootacar Mk I/II	197/9	1958-65	231	GB
Mk III Twin	324/13	1961-65	213	GB
Scoot-mobile	—/12	1946	274	US

This Scootacar Mk II was owned by the car's designer, Henry Brown, and features modified front bodywork effected by Brown himself.

Scorhill El Cid	602/31	1992-date	352	GB
[**SEAB**: see Flipper]				
SEAT 600	633/22	1957-63	322	E
Sebring Vanguard	Electric	1972-86	297	US
Shanghai Hai-San SW-710	297/13	1959-c80	—	CHI
Shelter	197-228/—	1954-58	—	NL
Sherwood	Electric	1981-?	—	AUS
Shibaura	—/—	1954	—	J
Shopper	—/—	1960s	216	S
Siata 500 Spyder	569/16.5	1948-49	—	I
Orchidea	569/—	1948	—	I
Amica	569/16-20	1949-55	—	I
Mitzi	398-434/11-12	1953-55	280	I
600/Amica	633/22	1955-60	370	I
500 Spider	479/21	1959-c60	370	I
[**Sicraf**: see P. Vallee]				
SIFTT Katar	652/35	1987-89	—	F
Sigma	652/—	1988	—	H
Sigmund	250/—	1950	—	D
SILA Autoretta	200/11	1960	—	I
Silaos Demoiselle	47-430/—	1985-88	230	F
Simca 5	569/13	1946-49	336	F
6	569/14-16	1947-56	—	F
Sinclair C5	Electric	1985	—	GB
SIOP Marathon	746/42-52	1953-55	360	F
Siva Parisienne	602/29	1970-75	—	GB
Smyk	350/15	1957-58	—	PL
SMZ	—/7.3-10	1956-70	—	SU
SZD	346/12	1971-75	—	SU
SND D50/D125/Onyx	325-654/6-12	1992-94	248	F
Sofravel	150/6.5	1948-49	270	F
SP Spi-Tri	Electric	1981-?	335	US
Spatz	191-250/10-14	1955-57	330	D
Spijkstaal	Electric	1984-86	209	NL
Stal	300/14	1958	328	PL
Staunau K 400	389/13	1950-51	412	D
Steinwinter Junior	50-125/—	1980	240	D
Stela	Electric	1941-44	—	F
Steyr-Puch 500	493-643/16-23	1957-73	297	A
650T/TR	660/20-40	1962-68	297	A
Haflinger	643/27	1958-75	314	A
126	643/25	1973-?	305	A
[**Stil**: see Voiture Electronique]				
Story	Electric	1941-44	—	NL
Stromboli	Electric	1989	285	CH
Stroj Flea Way	125/—	1949	220	F

Stuart	Electric	1961	—	US
Subaru 360	356/16-25	1958-71	300	J
Maia/K212	422/22	1960s	300	J
R-2	356/32-36	1970-72	300	J
Rex	356-544/28-31	1972-81	300	J
Rex	544-658/31-63	1981-date	330	J
Vivio	658/42-63	1992-date	330	J
Succes	—/—	1952	—	B
Sui Tong Rickshaw	—/—	1960s-80s	—	RC
Cub	—/—	1983	—	RC
Suminoe Flying Feather	350/12.5	1951-55	277	J
Sunrise Badal	198/10	1978-81	310	IND
Sunrise	—/—	1984	—	US
Super Kar	—/15	1946	—	US
Suranyi Typ 47	125/—	1946-47	—	H
Surrey '03	—/8	1958-59	—	US
Suzuki Suzulight	360/—	1955-57	299	J
Suzulight 360	360/21	1962-67	299	J
Fronte 360	360/25-36	1967-70	300	J
Fronte/Alto/Cervo	360-539/28-31	1970-82	300	J
Fronte/Alto/Cervo	539-543/28-40	1979-88	320	J
Fronte/Alto/Cervo	547-647/34-63	1988-date	330	J
Jimny	359-539/27-58	1970-date	300	J
Cappuccino	657/63	1992-date	330	J
Cara	657/64	1993-date	330	J
Wagon R	657/55	1993-date	330	J
CV-1	50/3.4	1981	190	J
T	578/24	1956-57	398	CH
Tama	Electric	1947-51	—	J
Tecoplan Leo	720/34	1988-90	250	D
Teilhol Citadine	Elec-123-325/—	1972-82	218	F
Messagette/Handicar	Electric	1975-c83	230	F
Simply/T50/T125	50-125/3	1981-84	213	F
325/400 TD	123-400/—	1984-86	217	F
Citadine 2	123-199/—	1985	—	F
TLX/25TXA	325-400/—	1986-90	246	F
Tangara	602/29	1987-90	350	F
Tempo Troll	197/7.5	1949-51	—	D
Thaon AT2	Electric	1975-76	150	CH
Tholome	345/14	1948-50	210	F
Thrif-T	—/10	1955	—	US
Tibicar Bella	123/5.5	1979-85	245	I
TiCi Mk 1	500/—	1969	183	GB
Mk 2	848/38	1972-73	226	GB
Tilbrooks	197/—	1953	—	AUS

Tilli Capton	—/15	1957	—	AUS
Tomcar	49-325/—	1983-85	246	F
Tomos Citroën 2CV/Dyane	425-602/13-31	1960-?	383	YU
Tourette	197/6-7.6	1956-57	259	GB
Towne Shopper	—/10.6	1948-49	—	US
Toyota Publica	697/28	1961-66	—	J
TP Kesling Yare	Electric	1979	427	US
Trabant P50	500-594/18-23	1958-64	—	DDR
601	594/23-26	1964-91	355	DDR
Tractavant	125-150/—	1951-52	265	F
Tria MOD 01	500-900/—	1990	260	I
Tri-Car Suburbanette	—/30	1955	—	US
Tri-Ped Microcar	50/—	1979	—	US
Trippel SK 10	498-600/18.5-26	1950-52	309	D
Tripod	50/—	1986	—	GB
Trishul	510/11	1982-c86	—	IND
Triver	339/14	1952-57	—	E
Trojan 200	198/10	1961-65	255	GB
Troll	663-748/30-33	1955-57	—	N
Twike	Electric	1993-date	—	CH
TZ Sider	296-350/12-18	1956-66	—	E
UMAP	435/24	1959	—	F
Urbanina	198-Electric	1964-73	183	I
Uttoro	250/6	1954	265	H
V2N Jet	125-200/10	1958-59	320	F
V200	200/12	1953	—	I
Vanclee Mungo/Rusler/ Emu/Emmet	602/29	1978-date	381	B/GB
[**Vanguard**: see Sebring Vanguard]				
Vannod	200/10	1958	—	F
Vaucelle Strop	125/—	1951-53	260	F
Vautrin	125-350/—	1951	265	F
VAZ 1111/1113 Oka	644-750/30-34	1986-date	321	SU
VEL	—/—	1947-48	—	F
Velam-Isetta	236/10.5	1955-57	238	F
Velorex	250-344/9-16	1954-71	—	CS
[**Veloto**: see Bel-Motors]				
Vespa	393/14	1957-61	285	F
Vessa Pilcar/Carville	Electric	1977-82	306	CH
VH	400/—	1961	—	E
[**Victoria**: see Spatz]				
Vignale Cherie/Gran Luce	633/22	1955	—	I
Gamine	499/18	1967-c69	302	I
[**Villeple**: see OTI]				

Vannod.

Vimp	197/—	1955	—	GB
Viotti 600	633/22	1955-56	—	I
Vitrex Riboud	47/2.4	1974-80	195	F
Addax/Gildax	47-50/2.6-4.5	1974-80	183	F
Garbo	49/4	1980	216	F
Vlah	500/9	1948	—	CS
Voisin Biscooter	125-197/6-9	1949-52	250	F
Voisin Lynx	652/36	1984	382	F
Voiture Electronique				
Porquerolle/Cab	Electric	1968-76	180	F
Volpe	124/6	1947-49	250	I
Volugrafo Bimbo	125/5	1946-48	238	I
Wasp	—/—	1984	—	AUS
[**Weidner**: see Condor]				
Weiss Pente 500	500/15	1946	340	H
Pente 600	600/18	1947-48	350	H
Wendax Aero WS 700	398/11.5	1949-50	370	D
Wendler 500 C	569/16.5	1952	—	I
Westfalia M50	125/—	1950s	—	D
Westinghouse Markette	Electric	1970s	—	US
WFM Fafik	148/6	1958	—	PL
Wiima	300/15	1957	—	SF
Wiles-Thomson	—/7	1949-52	—	AUS
Willam City/Farmer	123/5.6	1967-86	190	F
500	499/18	1972-73	220	F
Cyclo	49/—	1978-c80	—	F
Witkar	Electric	1970s	—	NL
[**Wolf**: see Kleine Wolf]				
Yamaha PTX-1	50/—	1983	—	J

Volugrafo (below) *and Volpe.*

Zagato 500	569/13-16.5	1947-49	—	I
750 MM	569-747/16-35	1951-52	—	I
600 Coupé	633/22	1955	—	I
Z 600 TS	633/22	1955	—	I
Zele 1000/2000	Electric	1974-c91	195	I
Hondina/Zanzara	354-499/22-36	1970s	—	I
Nuova Zele/Z-Car	Electric	1981-c91	258	I
Zaporozhets	400/—	1954	—	SU
Zastava 600	633/22	1956-60	332	YU
Zeda	350/12	1948	—	CS
[**Zeta**: see Lightburn]				
Zoe Zipper	—/—	1984	—	US/J
Zundapp Janus	248/14	1957-58	—	D
Pininfarina	600/30	1957	—	D
Zwickau P70	692/22	1955-59	374	DDR

Zoe Zipper.

By 1957, the Biscuter had become a national institution and was being offered in a sports coupé version called the 200 F.

The Sbarro-built Pilcar (later the Carville) was commercialised by the Swiss firm Vessa.

KEY TO COUNTRY ABBREVIATIONS

A	Austria	GB	Great Britain	RA	Argentina
AUS	Australia	GBG	Guernsey	RC	Taiwan
B	Belgium	GBM	Isle of Man	RCA	Central African Republic
BR	Brazil	GR	Greece	RCH	Chile
CDN	Canada	H	Hungary	RI	Indonesia
CH	Switzerland	I	Italy	RO	Romania
CHI	China	IND	India	RSM	San Marino
CI	Ivory Coast	IR	Iran	S	Sweden
CS	Czechoslovakia	IRL	Ireland	SF	Finland
D	Germany	J	Japan	SN	Senegal
DDR	East Germany	MA	Morocco	SU	Soviet Union
DK	Denmark	N	Norway	U	Uruguay
E	Spain	NL	Netherlands	US	United States of America
ET	Egypt	P	Portugal	VN	Vietnam
F	France	PG	Guinea-Bissau	YU	Yugoslavia
FL	Liechtenstein	PL	Poland		

OBSCURE MARQUES

THE following microcars are believed to have existed, but very little is known about them due to lack of documentation. If any reader has any information concerning these very rare and obscure cars, please contact the author via the publisher of this book.

Baldet Bluebird (GB, 1950s)
Bearcat
CAM
C-Donki
Claeys-Flandria Mobil-Car (B, 1970s)
Cuno Bistram
FIM
Gazelle (NL)
Hindustan Pingle (IND)
INAT (F)
Knudson
Lantrac
Marketeer
Minikin
Minilux
Motorette
Ngo
Petit Puce (F, 1970s)
SCC PO3
SVP Electronique (F)
Taylor-Dunn
UM
VEZ
Voiturette Jaegar (F)
Wattcar (F)
Westcoaster
XK-1
Zoe Little Giant/Runner (US)